The Temporality of Determinacy

Conor Husbands

The Temporality of Determinacy

Functional Relations in Metaphysics and Science

palgrave
macmillan

Conor Husbands
London, UK

ISBN 978-3-030-86529-0 ISBN 978-3-030-86530-6 (eBook)
https://doi.org/10.1007/978-3-030-86530-6

This Palgrave Macmillan imprint is published by the registered company Springer Nature Switzerland AG.
The registered company address is: Gewerbestrasse 11, 6330 Cham, Switzerland

Contents

1

Introduction

The way in which events in the world relate to the rules and laws which govern them has long been a matter of philosophical dogma. Both Ancient and modern thinking has embraced the uniformity and consistency of nature: the world does not fluctuate as if the caprices of a deity, but is orderly, causal, and intelligible to the mind. Prominent scholars have held that this intuitive precept requires exogenous rules and laws, acting on events from outside, quite separate from the systems they describe. Such rules and laws escape the ravages of time which affect these systems. They neither evolve nor change, no matter how profoundly the world metamorphoses: events occur in time, but laws stand outside of time, and the causes and forces which animate the world are the same for any epoch. Thus, some go so far as to label them "eternal," as if to mimic Spinoza's characterisation of substance. Nature evolves according to *laws independent of time*—and must, in order to be consistent. The alternative, it is claimed, would plunge the world into chaos, and unintelligibility to the human mind.

This commitment to the time-independence of laws has an important implication. It owes to the fact that laws are frequently taken to confer determinacy on events. If they are free from the dynamics of any given

C. Husbands, *The Temporality of Determinacy*,
https://doi.org/10.1007/978-3-030-86530-6_1

system, the determinacy they impart is atemporal. The subjection of an event to rules and laws is independent of time, and is in no sense a process. Events such as this unfold according to laws which are eternal, not evolving. In turn, if laws of nature are considered to be represented by functions—the language of modern science—the variables they describe are determined once and for all to be as they are at each moment. It is for this reason that modern authors speculate that the time variable *t* can be expunged from these functions which quantify this evolution over time, and that the truth or falsity of the corresponding laws cannot *become* such as they will be, once were or are at any time, but always are so. They have no temporal structure.

The main tenets of this line of thought can be condensed into the following statement—one operating covertly in a multitude of philosophical debates, in Ancient disputes about time and fate, in Hume's skirmish with Kant over the grounds of causality, and in present-day accounts of the foundations of science:

(I) Laws of nature are absolutely independent of functions of time. (Atemporality Conjecture)

The thesis of this book is the negation of this statement. It is not the case that the laws of nature are absolutely independent of time. Its argument for this conclusion adduces developments in the physical and non-physical sciences, as well as the commitments of certain philosophical schools, which force this atemporality conjecture to be jettisoned, and cast into relief a number of phenomena whose mode of determination—that is, their subjection to the rules of scientific theory—does, in fact, display a temporal structure. These phenomena conform to laws which evolve with them, and which, rather than acting from an unworldly eternity, belong also to the systems they make up.

Examples of this kind of time-dependency appear in diverse disciplines, from time-series analysis to international relations, from statistical mechanics to cryptography—all of which are drawn upon in the course of the argument. It is instructive to introduce a few in advance.

1. In behavioural economics, the bilateral dependency between markets and the expectations of participants in them produces recursive behaviour. The correct price for a given security depends on reciprocal expectations, which are formed and reformed over time.[1] This correct price is therefore projected and revised as expectations are adjusted, and is determined by the rules which evolve in step with the market: only after, for instance, large losses are incurred does plunging confidence reveal a security to have been valued incorrectly. In the research of Elena Esposito, this behaviour (which she labels "revisability") leads to a distinction between the order in which price movements occur and the order in which they are determined. As a result, the conditions under which price movements or volatility measures are determined cannot be described in a time-independent way.[2]
2. In the context of thermal physics, particularly statistical mechanics, macroscopic variables are described probabilistically. Ergodic behaviour, an important species of statistical-mechanical behaviour, such as that of mixing and Bernoulli systems in canonical formulations of the theory, exhibit probabilistic irrelevance: events sufficiently separate in time are not only correlated to a weak or decaying degree, but, in certain circumstances, entirely probabilistically independent of one another, so that $P(e_l | e_k) = P(e_l)$ for a pair of events $\{e_k, e_l\}$.[3] This fact is contrary to proponents of the atemporality conjecture, such as Russell, who contend that specifying a law of nature and an event at a given time determines all other events it describes, regardless of their position in time. Here, by contrast, events *become* determinate, as the system and the laws describing it evolve through time.
3. Brian Massumi's work considers the logic of threats posed by two sides in a military conflict. In the case of pre-emptive military action, this logic converts an external threat into a determinate enemy as a result of the pre-emptive attack.[4,5] The aggressor erroneously considers the threat to be the determining cause of the pre-emption. With this logic, there is a reciprocal, symmetric or even entirely inverted relationship between the determination of the pre-emption by the threat and that of the threat by the pre-emption.[6] The threat is determined as such *after* the act which pre-empts it. Thus, the determinacy of the events of the system, such as threats or pre-emptive acts, cannot be secured in advance

by a law: they are determined through time. For Massumi, the driving forces behind pre-emptive action are exogenous and, as such, no less temporal than the hostility and destruction which emerges from it.

These three examples, which span almost entirely unrelated fields, are a small sample from a much larger population of similarly time-dependent rules, laws and logics which violate the atemporality conjecture (in ways which will be described in much more precise terms later on). However counter-intuitive they may seem, their prominent role in their respective fields of research makes it impossible to uphold the atemporality of determination, for a broad class of deterministic approaches and assumptions.

This present work aims to expand upon this material, utilising it to explain and justify the notion of time-dependent laws as well as the abandonment of the atemporality conjecture. In particular, it provides arguments for rejecting this conjecture as a universal or absolute constraint on valid laws. The arguments involved rely on several theoretical registers, which divide the work into three parts, each of which looks at the specific challenges posed to the atemporality conjecture within a particular group of disciplines and sub-disciplines. At each stage, evidence is presented of the increasing limitations and mounting untenability of a time-independent construal of determination, that is, the way in which rules and laws fix the behaviour of the world, and the systems which make it up.

The section following this introductory chapter lays the groundwork for this thesis. The notions of causality and determinism are surveyed and defined more precisely. Canonical formulations of causal determinism are considered, such as those provided by the likes of Laplace, Popper and Russell, couched in terms of the predictability of systems' evolution or the uniformity of their laws, as well as alternative conceptions, such as that available in John Earman's seminal *A Primer on Determinism*, which provides a detailed contextualisation of different forms of determinism[7] using aspects of modern and classical physics.[8] Earman offers useful refutations of persistent myths—that classical physics provides a paradigm of deterministic behaviour, that stochastic processes exhibit behaviour which is essentially indeterministic, than determinism is equivalent to determinateness, i.a.. These help hone and focus the different senses of these terms discussed in this work. Second of all, the approach of Lindsay,

Margenau, Ushenko and other twentieth century commentators is considered. These accounts couch the determinacy of events in the disappearance of the parameter *t* from the fundamental dynamical equations of scientific theories, a disappearance which is said to express the intuition that these dynamics remain consistent over time, for systems closed to external influences. As will be seen, this account faces counter-examples from a number of directions. Thirdly, it considers the influential contribution of Bertrand Russell's writings. Russell attacks one misleading source of the relation between causal determinism and temporality, namely the asymmetry between cause and effect which, he shows, are absent from the descriptive mathematics. It is shown that his conception of functional relations[9] provides Russell 1918 a helpful framework for analysing temporality and determinacy in a broad range of contexts, not only physical science. Overall, this groundwork (i) enables a clearer definition of the atemporality conjecture, and (ii) more clearly distinguishes different species of determinism, determinateness, determinacy, and related terms. Analysed in the next chapter, they provide the foundation for the rest of the work.

The remainder of the work divides into three further chapters. Two of these three assess the impact of the physical and non-physical sciences on the atemporality conjecture, cataloguing examples drawn from scientific theories which admit time-dependent dynamics.

In the case of the non-physical sciences, the recent work in the philosophy of economics of Elena Esposito[10] is examined, and related to that of Suhail Malik, Elie Ayache, Jon Roffe, Donald MacKenzie and others. Esposito, in particular, develops a sophisticated analysis of how financial markets, whose dynamics depend on participants' inter-mixing expectations, obey a time-dynamic proper to them: time is internal to the markets, as a system of expectations. This bilateral dependency of market conditions on participants' expectations engenders a recursive temporal system.[11] Esposito's argument is reconstructed, and shown to depend chiefly on two insights: the internality or endogeneity of time-relations to the economic system being described, and a claim about the revisable and changing nature of the expectations commoditised by financial trades. It is shown that her work implies a distinction between the order of events taking place (such as transactions or valuations within an economic system) and the order in which they are determined. It is in this sense that her theory requires a commitment to the temporality of determinacy.

Another germane line of research, related to Esposito's, involves a number of innovative tools brought to bear on the analysis of financial markets in recent decades, as well as in other fields. The discipline of time-series analysis attempts to fit time-varying mathematical functions to data, such as commodity prices and equity valuations in quantitative finance, but also climactic data, disease infection rates, biological signals, and information from a range of other sources.[12] The functions of physics contemplated by Russell and others fix which events occur at different points in time (or, coterminously, the values or states explored by a system at different times); these explicitly time-dependent functions, supplemented with appropriate error terms, determine future and past data in terms of given conditions.[13] Models developed in order to describe stochastic processes, also common to the study of financial markets, consist of collections of random variables indexed, like time-series data, by time. For instance, a species of stochastic behaviour, Markov processes, exhibit analogous kinds of probabilistic irrelevance to some of the processes of statistical mechanics: the history of the variable and the way in which the past has generated the present are irrelevant.[14] These are adopted and adapted in models of stock market price fluctuations. Similarly, binomial models utilised for the modelling of financial options, interest rates and other quantities consist of sequences of Bernoulli trials—trials which (occupying the extreme end of the ergodic hierarchy) display not only probabilistic irrelevance but independence from trial to trial, each outcome being indeterminable by a time-independent rule or law. These models display a number of features which undermine necessary conditions for the validity of the atemporality conjecture—features such as serial correlation, the interdependence of time-periods from one another and the absence of correlations between variables at different lags, and the contextuality of measurements.[15]

Even more striking examples from the social sciences can be adduced with similar implications. The latter sections of the third chapter address the research of Brian Massumi and, in particular, his recent *Ontopower: War, Powers and the State of Perception*. Massumi finds in the logic of military interventions a number of startling features which force any potential application of the atemporality conjecture to be reconsidered. Putative causes of pre-emptive military action (a supposedly objective threat) are in fact generated by the pre-emption, and show up after this pre-emption by way of retaliation. Cause and effect, determining and determined, future and past are designations which end up being swapped as the dynamics of the system unfold.[16] The occluded, diffuse or intangible nature of the threats faced by a hegemonic power, whether insurgent or paramilitary forces, hostile nation-states who conduct espionage through cyber criminals, or other malicious actors, reflects no mere epistemic limitation on the part of the hegemony, but rather a more fundamental ontological indeterminacy. In the context of international affairs and international conflict (and the determinative forces involved in these studies), then, Massumi's work thus provides strong evidence of a logic active in international affairs—indeed one with incontrovertibly concrete ramifications—which defies the strictures of the atemporality conjecture.

Finally, the third chapter turns to an even more recent sequence of developments—this time in cryptography, a field with now well known applications in message encryption, cyber-security and the management of digital currencies. The relevant implications of this field for the present purposes of this work are, summarily, the following. Consider a system which involves agents transacting (exchanging goods and services for money) over a network. A reliable means to establish the order in which events occur is a necessary condition for the viability of these systems. Whether goods were sent before or after payment was received, or whether the contents of an individual's bank account were emptied prior to attempting to pay someone from them—such questions have obvious implications for the integrity of whatever economy rests on them. The cryptographic algorithms employed by these systems, exposited in more detail in the third chapter, determine this order with an idiosyncratic procedure which achieves distributed consensus. The upshot of this procedure is that the determination of the system's ordinal time is itself a process, which requires computational power to be expended, nodes to interact, and agreement, couched in the outcome of cryptographic functions, to be reached. The process implies a distinction between the order of these events and the order in which they are determined, a familiar mark of temporal determinacy.

The fourth chapter of this work concerns itself with the evidence for temporal determinacy provided by the physical sciences. It opens with a review of recent research in the context of cosmology, such as that of Lee Smolin, whose theory of cosmological natural selection provides a framework for the variation of parameters in fundamental physical laws over time. Subsequently, it explores the discipline of statistical mechanics, frequently implicated in the debates over the philosophy of time and causality. The temporally asymmetric behaviour which appears to be its core subject matter, often attributed to the second law of thermodynamics, which concerns entropic increase, has frequently been regarded as a significant conundrum in academic discussions of the topic.[17] This section introduces the framework of dynamical systems theory often used for theorization about the discipline's foundations, and uses this framework to explore the relevant implications for the argument of ergodic systems, and different sub-classes of ergodic systems (such as mixing and Bernoulli systems). Summarily, it is argued that the notion of probabilistic irrelevance which arises in the context of these systems effectively undermines the atemporality conjecture, in spite of the fact that systems remain deterministic in an important sense.[18] Using the research of Charlotte Werndl and others, it is also shown how similar features also obtain in the domain of chaotic systems.[19] The temporality of determinacy is thus borne out by the fact that events can be independent of past events, independent of sufficiently far past events, or entirely independent, depending on the tier of the ergodic hierarchy they occupy.

Moving away from the context of thermal physics, statistical mechanics and chaos, the chapter proceeds to the theory of quantum mechanics, responsible for contributing an abundance of counter-intuitive results for philosophers concerned with causality, determinism and temporality. It discusses claims of retrocausal behaviour (sometimes referred to more loosely as backwards causation[20]) on the part of Huw Price, expatiating as he does the view that realism and time-symmetry together weakly imply that this kind of behaviour must exist, and its rigorization by Matthew Pusey and Matthew Leifer. These claims establish a specific sense in which the denial of retrocausality is incompatible with time-symmetry, subject to a number of plausible assumptions.[21] The chapter then considers the controversial transactional interpretation of quantum mechanics, as well as the responses of Tim Maudlin, Feynman and Wheeler. The section concludes that, whilst no incontrovertible aspect of quantum mechanics—whether the formalism or its experimental results—of itself entails the temporality of determinacy, many of the most prominent candidate interpretations of the theory are either consistent with or suggestive of such a perspective, and further weaken the atemporality conjecture.

This text's fifth chapter turns to philosophy, with a particular focus on the work of Kant. This chapter argues that Kant's metaphysics, as exists in the *Critique of Pure Reason, Critique of Judgement, Prolegomena*, and other tracts, amount to an important juncture in the development of thinking about determinism and time. Kant's contributions in this arena are, of course, enduring and exert a commanding influence today. But there are several unique aspects of his work which make it indispensible source of material for this argument as well. First of all, Kant's response to Hume's associative deflation of causality attempts to salvage the now-familiar commonsensical intuitions about the uniformity and consistency of the natural world, in contrast to the rhapsodic and changeable one at alleged stake in Hume's vision. This fact supports the claim (put forward also by Lindsay and Margenau) that these characteristics are definitive of causal determinism.[22] These questions of uniformity and consistency are in this sense central to his philosophy. Secondly, and more importantly, several theses defended by Kant in his critical works bear directly upon the issue of the subjection of events to natural laws. The doctrine of the unity of apperception posits a connection of all possible representations according to laws, performed by inner sense,[23] and specifies what Kant takes to be a necessary condition for preventing an anarchic rhapsody of perceptions.[24] Further, the Analogies of Experience contain some of Kant's most explicit work on the subjection of events to rules, such as his thesis that everything that happens requires another event from which it follows according to a rule.[25] Finally, the schematism of the understanding, which demonstrates the applicability of the categories of the understanding (the subject of the transcendental deduction) to the manifold of intuition and thus the possibility of subsumption of intuitions under these categories, considers the way in which these intuitions, given in time, interact with the universal and necessary rules which are not. The schemata corresponding to the categories, amounting to *a priori* determinations of time in accordance with rules,[26] therefore supply vital evidence as to whether or not the atemporality conjecture stands or falls in the context of Kant's work. Thirdly, the theory produced by Kant in his later *Critique of Judgement*, which asserts nature's conformity to principles of purposiveness and systematicity, explicitly temporalizes these rules by describing their formation by the mind as a process and activity.

In this way, the fifth chapter argues for two conclusions. First of all, the first *Critique* contains an important distinction between temporality of determinacy and indeterminism, conceived as the inconstancy of nature. This in turn enables time dependent physical laws to be reconciled with the uniformity and consistency of nature. This, as we have already seen, is an important intuition which appears in causally deterministic principles, as well as a strong commitment of Kant's.[27] Secondly, Kant's conceptual arsenal permits the conciliation of a robustly deterministic metaphysics, endorsing the universality and necessity of the laws of nature (including particular causal laws), with temporal determinacy. It thus lends force to the argument that the principles of consistency and uniformity do *not* entail time-independence of the subject laws. The viability of this approach is then assessed in light of recent commentary on Kant's philosophy of causality such as that of Watkins, Schaper, Ginsborg, Guyer, Walker, Teufel and Bader, as well as Heidegger and Hegel.

The arguments summarised here, then, yield a case against the atemporality conjecture. The question as to the temporal status of events' subjection to rules and laws surfaces in many periods, and follows a subterranean trajectory through the history of philosophical thought. The assumption that the uniformity and consistency of nature, a capstone of determinism, entails the atemporality of its laws, is rarely addressed. Yet it lies uninterrogated at the heart of philosophical and methodological approaches to the sciences. From the idea that variables (appearing in predictive and explanatory models) have no fundamental time-dependencies, or that their essential and functional relationships are preserved with variation over time, to the assumption that there is no temporality and, *a fortiori*, no order or process by which events become determined,[28] to the idea that the past is devoid of causal efficacy—whether avowed as such or not, all of these stand in close logical relations to the atemporality conjecture.

However counter-intuitive it may seem, there are more and more grounds for abandoning this assumption. Instead, this book argues for the temporality of determinacy as a plausible component of a broadly deterministic metaphysics, one demanded by accelerating developments in the social and physical sciences, and one which can be reconciled with relevant aspects of the philosophical canon. Time does not only order events in time, quantify the duration of events in time, or parameterise the passing of events. It inhabits the dynamics themselves, and introduces variability into the functions which express them. The presence of time as a variable in these equations is no aberration to be expunged, nor a signifier of a theory which is improperly formed, as some commentators have suggested. To the extent that these functions determine the evolution of the world, enabling prediction and retrodiction to be carried out and explanations to be formulated by an observer, they stand outside of time no more than do the events which are their arguments. The time-dependency cannot be straightforwardly extirpated. The following chapters chronicle the increasing volume of material which supports these theses.

Before turning to this task, it is instructive to examine how this material relates to traditional debates in the philosophy of time likely more familiar to the reader. The time frame in which different perspectives on the temporality of determinacy have arisen has certainly seen the emergence of a large number of problems in this area. Going back sufficiently far, the Ancients' discussions of bivalence related to the question of whether propositions' truth values could vary over time, with some using this principle to argue for eternal causes. In modern philosophy, the relationship between logic and the metaphysics of time takes on other forms, with particular discussion of the question as to how to represent change in logical form. Kant's canonical description of alteration distinguishes between substances which are maintained throughout this alteration and its accidents, viz. particular determinations of substances, which, by contrast, fluctuate. In the *Critique of Pure Reason*, Kant considers the being and not-being of a particular determination of a substance, one of which succeeds the other in appearances, to constitute change.[29] Simply, an instance of change or alteration can be presented in the following predicate-subject form: for t_1, Pa; for t_2, $\neg Pa$ where these two times t_1, t_2 succeed one another. The subject-predicate form and the distinction between substance and accident inclines one's thinking to a certain view of change—one according to which change is something which happens to entities which pre and post-exist the change, and remain unchanged throughout the change.

This distinction has been controversial and has incited a number of sceptical attacks. Nietzsche, particularly in his later writings, consistently inveighs against the opposition of subject and predicate at the heart of this view of change. Hume substitutes in place of the subject persisting through change a bundle of perceptions standing in associative relations but without substratum. These strains of scepticism are only compounded by the prominence of process philosophy and the thinking of twentieth century authors such as Henri Bergson and A.N. Whitehead. In the case of the latter, a polymath of mathematics, philosophy and logic, a metaphysics is created in which entities are replaced by processes, drops of perception, or occasions. Over time, prehensions, the basic units of such processes, which Whitehead considers to be occasions, grow together into concrete experiences. Thus, his *Process and Reality*, he explicitly draws a distinction between this growing, which he terms concrescence, and on the other hand the transition from particular existent to particular existent.[30] With processes becoming elementary, rather than the differences between determinations of substances in successive moments, change is no longer reducible to temporal variation of the form t_1: Pa; t_2: $\neg Pa$.

Just as the logical puzzles which arise in the philosophy of time are numerous and go beyond the proper way to represent change, so could the same be said of its metaphysical aspects. The controversy as to the absolute or relative nature of time is a traditional example—a controversy which intensified rapidly in the wake of Newtonian physics. Newton, in his *Principia Mathematica*, talks of distinguishing between true and apparent motion as the motivating principle of the work, and envisaged absolute time as one of the core principles through which the two could be established. He conceived elaborate thought experiments, such as the bucket argument, in order to refute the pretences of relational theories to constitute the same degree of explanatory power as absolute space, but with a leaner, more parsimonious ontology. This legacy was a fecund source of discussion, culminating in the famous exchanges between Leibniz and the Newtonian Samuel Clarke. Clarke defended time (as he did space) as an entity independent of events in time, a move Leibniz rejected on diverse grounds, in favour of relationism: time is dependent on events in time, and in no way contains or precedes them; thus, empty time, time free of events, cannot be. Such quarrels between absolute, substantivalist views of space and time and relationism have persisted into the post-Newtonian epoch, with defences of and attacks on manifold substantivalism in the context of general relativity.

In the twentieth century, McTaggart's famous 1908 paper "The Unreality of Time" considers whether change can be reduced to a sequence of ordering relations between moments: $M < N$, $N < O$, … The basis for this reduction involves a distinction between different time-series, A and B.[31] The B-series, for McTaggart, consists of such a sequence, effectively relations of being earlier or later than the other moment. These relations are static and do not change; propositions about them are true or false at any and every time—reminiscent of the "block universe" world-picture which gained popularity in the wake of special relativistic physics. The A-series supplements this series with a designation of which moment of these is the present. Varying designations of moments as past and future follow from this move, with all moments occurring prior to the present being deemed past and all later ones future. Why introduce this distinction? McTaggart's purpose is to construct a *modus tollens* argument: he demonstrates that the reality of time requires both A-series and B-series, and yet the A-series is contradictory and cannot exist. Passage or transition *from* one moment *to* another might intuitively be regarded as our everyday experience of time, with one moment of all moments being deemed the present. Static relations between moments cannot, for McTaggart, by themselves constitute this experience of real time. However, this A-series is itself incoherent. It requires the predicates "past," "present" and "future" be assigned to each moment in order for time to flow, but these predicates are incompossible. This, McTaggart posits, entails the unreality of time. From this vantage point, processes and changes cannot be reduced to static ordering relations between moments.

The atemporality conjecture and the philosophy motivating it are not reducible to a conjunction of one or another position with respect to these various debates. Accepting or rejecting the atemporality conjecture, and the reasons for doing so articulated in this book, need not entail any particular stance with respect to these debates in the philosophy of time; nor is the temporality of determinacy an issue which arises from them directly. Nonetheless, there are conceptual relations between them which are worthy of study, providing as they do useful context for the temporality of determinacy, and exhibiting its connections to more familiar questions, as well as the consequences of endorsing it. The conjecture asserts that laws of nature, expressed in the form of mathematical functions, are absolutely independent of time. Thus, rather than the logical form of change itself, the temporality of determinacy problematizes the engine of change, that is, the laws which encode the dynamics which impel these changes, and determine past and future states. The question as to the time-dependency of these laws thus concerns how the laws governing change can be represented. As future chapters will show, once the various forms of the atemporality conjecture are more fully disinterred, a number of questions arise as to their form, from which variables should or should not appear in the dynamical equations to their responses to various coordinate transformations. These are, so to speak, the equivalent of the traditional questions as to whether the logical form of change is amenable to description by subjects and predicates.

Where does the atemporality conjecture stand in comparison to the absolutist and relationalist approaches to time? Strictly speaking, whether or not laws are independent of time says little in and of itself about whether time is a nothing more than a relation between events, or entirely independent of them. However, there are loose connections between the two, which are stimulated all the more by recent science. The relativistic physics of the early twentieth century, constituted in Einstein's special and general theories of relativity, reframes measurements of time: these change depending on the reference frame or, from a dynamical perspective, the relative state of motion of the observer. In special relativity, moving clocks run slower and travelling bodies contract. With general relativity, once gravity is introduced into the mix, fields also distort such measurements, and transformations are required in order to reconcile measurements made in different frames. Superficially, this might be thought to incline the relativistic physicist away from an absolutism with respect to time. Einstein's principle of general covariance, which later chapters will elaborate on, states more or less that laws must apply to systems whatever their state of motion, and should be describable by equations which are valid for all co-ordinate systems. This too might incline one to believe that they should be time-independent, at least in as much as time is considered to be an aspect of such a co-ordinate system. In some cases, commentators on the temporal status of laws have seen that there is indeed a close relationship between developments such as these and the temporality of laws: Lee Smolin, for instance, puts this question at the heart of his *Time Reborn*. He claims that laws are not timeless and that they can and must evolve with the systems that they describe.[32] Accordingly, for Smolin, time takes on a metaphysical standing above and beyond relations between events in time, by parameterising the evolution of the dynamics of systems. In turn, Smolin's perspective weakens the connection between relationalism and time-dependent laws.

Similar connections can be drawn between the temporality of laws and the questions raised by McTaggart. The minimalistic B-serial view of time distinguished in his paper involves only an ordered series of moments in relations of being before and after one another. Just as McTaggart argues that such a view is insufficient for the reality of time, failing as it does to account for passage or real change, so might one suspect it to struggle to account for time-dependency in the laws of nature. Whatever form this dependency ultimately takes, time belongs not only to events (occurring at the moments in the B-series) but also to the laws which determine them. The temporality of determinacy aligns in this sense with McTaggart's instinct that abstract relations between moments or events do not capture all of time's relevant metaphysical content—content future chapters will seek to expand upon in detail in the course of arguing against the claim that laws are absolutely time-independent, and developing their own counter-thesis.

Notes

1. Esposito (2011), pp. 27–28.
2. Brassier et al. (2017).
3. Berkovitz et al. (2006), pp. 668–671.
4. Avanessian and Malik (n.d.).
5. Massumi (2015), p. 15.
6. Massumi (2015), p. 100.
7. Earman (1986).
8. Earman (2004a), p. 1.
9. Russell (1918), pp. 194–195.
10. Esposito (2011, 2018b).
11. Esposito (2011), pp. 24–25.
12. Indeed, time-series analysis is not confined to non-physical sciences.
13. Karimi (2010), pp. 2–3.
14. Hull (2002), p. 216.
15. Abbott (2001), pp. 51–59.
16. Massumi (2015), pp. 6, 117.
17. Callender (2016b).
18. Earman (1986), pp. 16–18.
19. Werndl (2009a).
20. Price (2001).
21. Price (2012).
22. Kant (1933), B196/A157.
23. Kant (1933), A108.
24. Kant (1933), B196 / A157.
25. Kant (1933), A190.
26. Kant (1933), B184/A145.
27. Kant (1933), A833/B861.
28. This is a position which appears frequently in Pierre Klossowski's philosophy of time. A more specific formalisation and analysis is given in Husbands (2020b), p. 131.
29. Kant (1933), B233. Elsewhere (A187/B231) Kant alludes to alterations as ways of existing of a given object followed by other distinct ways of existing of this object.
30. Whitehead (1985), p. 210.
31. Another, less widely discussed series called the C-series, is also distinguished.
32. Smolin (2013), p. xiv.

Bibliography

Abbott, A. (2001). *Time Matters: On Theory and Method*. University of Chicago Press.
Avanessian, A., & Malik, S. (n.d.). The Time-Complex. *DIS Magazine*. Retrieved September 26, 2019, from http://dismagazine.com/discussion/81924/the-time-complex-postcontemporary/
Berkovitz, J., Frigg, R., & Kronz, F. (2006). The Ergodic Hierarchy, Randomness and Hamiltonian Chaos. *Studies in History and Philosophy of Modern Physics, 37*, 661–691.
Brassier, R., et al. (2017). Pricing Time: Outline and Discussion on Suhail Malik's "The Ontology of Finance". *Identities: Journal of Politics, Gender and Culture, 14*(1), 84–128.
Callender, C. (2016b). Thermodynamic Asymmetry in Time. In E. Zalta (Ed.), *The Stanford Encyclopedia of Philosophy*. Retrieved April 12, 2020, from https://plato.stanford.edu/archives/win2016/entries/time-thermo/
Earman, J. (1986). *A Primer on Determinism*. D. Reidel.
Earman, J. (2004a). Determinism: What We Have Learned and What We Still Don't Know. Retrieved April 10, 2020, from https://www.ucl.ac.uk/~uctytho/Determinism.pdf
Esposito, E. (2011). *The Future of Futures: The Time of Money in Financing and Society* (E. Esposito, Trans.). Edward Elgar.
Esposito, E. (2018b). The Time of Money in Finance and U.S. Society. *Finance and Society, 4*(1), 15–25.
Hull, J. (2002). *Options, Futures and Other Derivatives* (5th ed.). Prentice Hall.
Husbands, C. (2020b). *Klossowski's Semiotic of Intensity*. De Gruyter.
Kant, I. (1933). *Immanuel Kant's Critique of Pure Reason* (N. Smith, Trans.). Macmillan.
Karimi, K. (2010). *A Brief Introduction to Temporality and Causality*. Retrieved April 11, 2020, from https://arxiv.org/ftp/arxiv/papers/1007/1007.2449.pdf
Massumi, B. (2015). *Ontopower: War, Powers, and the State of Perception*. Duke University Press.
Price, H. (2001). Backwards Causation, Hidden Variables and the Meaning of Completeness. *Pramana, 56*(2, 3), 199–209.

Price, H. (2012). Does Time-Symmetry Imply Retrocausality? How the Quantum World Says "Maybe". *Studies in the History and Philosophy of Modern Physics, 43*, 75–83.

Russell, B. (1918). *Mysticism and Logic and Other Essays*. Longmans, Green and Co.

Smolin, L. (2013). *Time Reborn*. Penguin.

Werndl, C. (2009a). What Are the New Implications of Chaos for Unpredictability? *The British Journal for the Philosophy of Science, 60*(1), 195–220.

Whitehead, A. (1985). *Process and Reality*. Free Press.

2

Determinacy and Functional Relations

Introduction

John Earman's seminal *A Primer on Determinism* provides an encyclopaedic survey of the role of deterministic assumptions within physics, its purview encompassing pillars of the field from Newtonian mechanics through to Special and General Relativity, Chaos Theory and Cosmology. In his introduction, he identifies an impasse which dogs any philosophical analysis of determinism. On the one hand, Earman suggests, the bewildering array of competing and contrasting characterisations of the doctrine naturally suggest the need for its firm reanchoring in the concrete results of scientific research and experimentation, as if to counter unbridled metaphysical conjecture with empiricism. On the other hand, some prior understanding of the very same doctrine would seem to be required, not only to ascertain the implications of these concrete results for determinism, but also, for those who take determinism as a condition of possibility of scientific investigation, to derive and formulate these results in the first place. The problem Earman identifies calls to mind Meno's paradox from the eponymous Platonic dialogue: how can one search for answer without first knowing its nature?[1] Earman's begrudging conclusion is that no definition of determinism that is both precise and general can be formulated, leading him to substitute in its place an indefinite one instead.[2]

C. Husbands, *The Temporality of Determinacy*,
https://doi.org/10.1007/978-3-030-86530-6_2

This conclusion is only reinforced by the text itself which, despite confining its focus to physics, sees countless definitions jostle for position. Taking leave from his broad distinction between epistemological and ontological formulations of the doctrine in his paper *Determinism: What We Have Learned and What We Still Don't Know*,[3] the introduction alone of *A Primer on Determinism* contains a litany of further species which emerge within each. The well known definitions of Laplace, Karl Popper, and Bertrand Russell, for instance, are all subsumed under the concept of predictability, being a mark of epistemological rather than ontological determinism. In subsequent chapters, further complexity emerges. Dispensing with the myth of classical physics as quintessentially deterministic, charting unstable systems whose compliance with deterministic principles differs from scale to scale, legitimating randomness, stochasticity and chance as admissible modes of deterministic behaviour—all of these add nuance to the role of determinism in physics. Accordingly, Earman makes no attempt to contrive a single definition from this discrepant picture, presenting what he deems modestly only an incomplete sketch.[4]

These observations betoken the vastly complicated and diffuse landscape of competing interpretations of determinism. It indicates that determinism is more a category or umbrella term than a unitary principle. Straying outside of Earman's domain, from physics into other physical sciences, from the physical sciences to the non-physical sciences, or from philosophy of science to metaphysics, inflates this complexity by orders of magnitude. This fact only increases the urgency of this present chapter's task: to define the most important terms and concepts which form the architecture for the argument of the ensuing chapters. The atemporality conjecture alluded to in the introduction can be articulated and a trial with it made only if the notions in terms of which it is couched are themselves spelt out clearly—just as reconciling determinism with the refutation of this atemporality conjecture requires clarity on the meaning of this term. Despite their importance, none of these are subject to any more of a consensus among metaphysicians or philosophers of science than determinism is within Earman's account of it—which is a nuanced one, even without paying attention to the additional complications which appear once one considers the status of determinism outside of physics, in disciplines such as neuroscience, theology or economics. This chapter therefore aims to clear up these matters, and to set out what is meant by the terms deployed throughout the rest of the book.

It should be noted that there are several of these terms, and that determinism is not the only equivocal concept involved. Any statement using the words "temporal" or "atemporal" stands or falls with a clear view of what these mean, too. Determinacy, understood as the determinacy of events subject to laws of nature, in turn requires a conception of what events and laws mean. Events and laws, in turn, relate closely to discussions of causality, such as Spinoza's or Kant's, as well as in those of modern scholars such as Hoefer and Mellor.[5,6] Consider, further, the notion of a physical law, or a law of nature, in the context of the physical sciences. Historical and contemporary works in the philosophy of science frequently refer to the discovery of laws as being central to the purpose of science, and to the universal relations that they encode as if to underwrite science's claims to possess a special kind of truth. In Hempel's Deductive-Nomological model, laws appear as components necessary for science's ability not only to predict and retrodict but to *explain* phenomena.[7] In Imre Lakatos' approach to understanding the structure of scientific theories, laws often inhabit their *hard core*, consisting of foundational and indispensable theoretical assumptions—such as Newton's laws of classical mechanics—which combine with secondary *auxiliary hypotheses* and flexible working assumptions to predict and explain experimental results. In Kant's metaphysics, laws are not only definitive of scientific enquiry, but essential to the rationality of nature itself, in that a rational doctrine of nature can constitute natural science only if its distinctive laws are cognized and grasped. For Kant, nature itself means law.[8]

There is thus some consensus (albeit not universal[9]) as to the importance of laws for scientific research. Equally, though, there is divergence as to what lawlike behaviour consists in or how laws ought to be defined. Scholars posit supervenience on facts and observations, furnishing relations between universals, signification of metaphysical, nomological or other necessity and many other incompatible characteristics as the core properties of laws. In some cases, they posit the vacuity of the debate itself.[10] Nor is there any more consistency to be found in the history of the concept, looking to the work of its progenitors. Edgar Zilsel's fastidious account of the genesis and evolution of the notion of physical laws, in his compendium *The Social Origins of Modern Science*, offers a useful illustration, showing how the notion traces out a malcoordinated path through the thought of Galileo, Kepler, Pascal, Descartes, Spinoza and Newton, impregnated with juridical resonances and traces of theological determinism, and with only vague and variable connections to its contemporary equivalents of consistency and universality.[11] For example, for Zilsel, the modern idea of relating events subject to laws to one another through mathematical functions first arose clearly with Galileo; yet Galileo very rarely spoke of laws with any of the physical meaning common today. Instead, *functional relations*, relations which connect physical quantities to one another, were to be his object (and the object of science), quite separate from the references to laws in the vernacular of theologians.[12] Like determinism, then, the notion of physical law, which is essential to the course of this book, is multi-faceted. It poses ambiguities, and lacks a straightforward scholarly consensus.

A similar fate befalls that of causality. Subject to unwieldy complications in meaning and inconsistencies in usage, Russell declares it misleading, confusing, and worth expunging completely from the lexicon of philosophy.[13] These problems hardly appear to have abated but rather to have inflated in the decades since Russell's protestation. Indeed, Earman also emphasises the neglected ambiguities latent in familiar formulae such as the following:

(A) Every event has a cause;
(B) Every event has a cause; like causes always produce like effects.[14]

He notes how the terms "causation," "causality," and "cause," are taken to be unproblematically cognate, in spite of these complexities. Similarly, Miller, in his paper "The Principle of Causality," analyses tautological, empirical and imperative conceptions of "the" principle, distinguishing a number of different approaches.[15]

What conclusion ought to be drawn from these observations? What do these presentations of determinism, law, and causality have in common? Summarily: No consensus exists as to the meaning of these key terms on which any discussion of the temporality of determinacy will find itself deeply reliant; nor are there any immediately ready-to-hand definitions. This chapter's mission to provide a clear, working basis for them is therefore crucial groundwork. For the sake of precision, they can be listed:

- Determinism
- Determinacy
- (A)Temporality
- Laws, laws of nature
- Causation, causality

Thus, not only determinism, whose ambiguities are revealed by Earman and others, but *relata* such as temporality, determinacy, law, rule, causality and causation all remain to be clarified. Doing so requires first of all a degree of acclimatization to the context of these competing formulations. This is provided in the first section of this chapter, which charts the evolution of functional dependencies as a framework for understanding laws. The chapter then considers the various formal tools available to define determinism, determinacy, and what is meant by the temporality or atemporality of these laws. First of all, these tools allow different positions in the debate over the temporality of determinacy to be formulated precisely, giving a plainer view of the stakes involved in this project. Second of all, they facilitate inter-disciplinary comparisons. These will prove fruitful once the discussion moves, as it does in subsequent chapters, to the developments in the physical and social sciences which impinge upon them. Finally, with the aid of these tools, this chapter sets out and analyses what it styles as the "atemporality conjecture," drawing upon the work of Robert Lindsay, Henry Margenau, Bertrand Russell and Herbert Feigl.

Causality, Determinism and Predictability

The image of Laplace's demon is among the canonical illustrations of determinism, and a helpful starting point for any analysis of causality and determinacy.[16] Laplace envisages an intelligent creature who wields an exceptionless grasp of all natural forces and of all beings subject to them, speculating that such a grasp would enable knowledge of the world's entire evolution—future and past—and the wholesale extrusion of uncertainty from its view of nature. This passage, cited in innumerable academic discussions, displays many of the most familiar elements of this doctrine as well as many of its inadequacies. Dividing the universe into the beings which comprise it and the laws which govern their behaviour, Laplace holds that instantaneous knowledge of the state of these beings, together with knowledge of the laws, suffice for universal prediction and retrodiction of the universe's evolution. The appearance in his statement of categories such as intelligence, comprehension, and certainty manifests an epistemological turn: it fixates less on the evolution of the world than on what can be known about it.[17] On this basis, modern scholars have taken Laplace to be one of the pre-eminent figures responsible for casting determinism in terms of predictability—although, in symmetrical fashion, Laplace's work itself harbours vestiges of forerunners such as Augustine, Aquinas, Calvin and Leibniz who, working in the tradition of theological determinism, consistently couched their discussions in terms of the foreknowledge of an infinite intelligence, albeit divine as opposed to demonic.[18]

In spite of recent doubts about the predictability criterion, it has been prevalent in philosophical literature on determinism. As we have already seen, Earman locates it in influential accounts published more than a Century after Laplace's essay. It surfaces in the writings of Carnap, Max Born, van Kampen and Leon Brillouin.[19,20] And, in one of the most noteworthy examples on the part of a prominent philosopher of science, Popper subscribes to the imagery of Laplace's demon, qualifying it only with the proviso that predictions valid with arbitrary precision require sufficient precision as to the demon's grasp of initial conditions.[21] The finitude of the demon's predictive powers thus originates in the partial, not complete, available description of the states whose evolution is in question.[22]

Granted the predictability account's influence, what are the implications for the aims of this chapter, namely to contextualise and elucidate the concepts relied upon for the arguments of later chapters? The rationale for the definitions of these concepts are cast into clearer relief by a clear understanding of the demerits of the predictability approach, responsible for its now waning command over the main debates in the metaphysics of determinism. It is therefore beneficial to briefly examine these demerits. Doing so will make clear why the simple, general framing of determinism in terms of *functional relations* proves a more robust alternative, which penetrates the many different contexts in which physical and non-physical laws merit discussion. First of all, counter-examples to the thesis that the predictability of systems is necessary and sufficient to make them deterministic are readily forthcoming, e.g. in both modern and classical physics. Second of all, as Earman and others have emphasised, the predictability criterion involves a conflation of the ontological with the epistemological considerations at stake in these debates. Thirdly, and most decisively, predictability puts a temporal asymmetry into determinism which is unmotivated, and at odds with the structure of many deterministic theories.

It is helpful to look at a number of examples of these defects. Once again, it only makes clearer the need to substitute in place of predictability and causality the framework of functional relations

(I) Research in the field of chaos theory is an evolving testament to the class of deterministic systems whose behaviour is not predictable. Chaotic systems, which inhabit fields as diverse as biology, climatology, finance and physics, are often characterised in terms of their asymptotic unpredictability, in which imprecision in the initial state of the system means that combinations of initial conditions, no matter how adjacent they may be, eventually diverge enough to usurp predictability.[23] Not only this: as later chapters will discuss in more detail, such chaotic systems display the further characteristic of probabilistic irrelevance; the probabilities of occurrence of states of the system separated by sufficient lengths of time are independent.[24] Counter-intuitively, even within the traditionally deterministic environment of classical mechanical systems, such as Newtonian systems, examples of unpredictability are prevalent. In his essay "Causation as Folk Science," John Norton envisages a spherical mass sitting at rest atop a frictionless hemisphere, and claims that remaining at rest atop the hemisphere or rolling down the dome in a number of directions are dynamical trajectories equally consistent with the system's laws.[25] More recently, Charlotte Werndl's research establishes the possibility that certain types of indeterministic descriptions simulate deterministic descriptions, in the sense that they yield identical predictions and entail observational equivalence, a possibility which further undermines the link between predictability and determinism.[26]

(II) Second, turning to the predictability criterion's conflation of the ontological with the epistemological questions of deterministic metaphysics already outlined, Earman's treatment is particularly unfavourable and provides a useful, compact illustration of what he takes to be its primary limitation. The criterion, he claims, is ultimately stated in terms of constraints on our knowledge, despite masquerading as an ontological principle, inducing needless confusion and ambiguity and misleading scholars.[27] His accusations are followed by Primas,[28] Werndl,[29] Hoefer,[30] and others.

(III) Aside from its questionable integration of ontological and epistemological perspectives on the problem of determinism, the predictability criterion suffers from an additional weakness: its implicit temporal asymmetry. Predictions are made in respect of future events. For a set of initial conditions given at a time t_i, equivalent to Laplace's references to the specific situation of the beings and things composing nature, the *subsequent* conditions, those at $t_j > t_i$, are predictable. A system's conformity to deterministic dynamics or refusal of them thus consists in the truth or falsity of statements made about its future, not about its past. The predictability criterion thus exhibits directionality. Yet, intuitively, determinism does not seem to be a doctrine with directional content. Neither James nor Earman nor Russell nor any other leading contributors to the philosophical discussion cast it in these terms. Moreover, not only do the laws which many physicists take to express the fundamental dynamics of a number of theories fail to exhibit any such directionality; on the contrary, in many cases, they actively refute it, to the extent that they exhibit time-reversal invariance.[31] This holds, for instance, of the laws governing many of the fundamental interactions—the Schrodinger equation in quantum mechanics, for instance, or Newton's laws. Thus, Hoefer declares that for a very wide range of theories which incorporate legitimate laws of nature, determinism logically implies *bi-directional* determinism, meaning that a suitable definition of the world-state at a particular time, coupled with the laws of such a theory, determines the world-state before and after. The very objects of the verb "determine" are both past and future.[32] Summarily, predictability is an asymmetric relation, yet the dynamics which are the best candidates for defining what it is to be a deterministic theory exhibit temporal symmetry.

There is some irony in the fact that nobody demonstrated a clearer grasp of these symmetries than Laplace himself. Despite the asymmetry involved in the predictability criterion, and despite the frequent imputation of this criterion to him above all others, Laplace's clear reference to the possibility that his demon might grasp in one formula the *past* as well as the future of the world provides clear evidence that he knew well the demon's vision would penetrate as far backwards in time as it would forwards. Given appropriate data as to the state of nature's constituents—in modern terms, a set of initial conditions—and a formula for the "movements" of the world, temporally distant events in either direction in time can be ascertained with certainty. There is thus no directional privilege at work in the chain of events making up this universe, but rather a temporal symmetry at work in the core of his *Gedankenexperiment*. This passage from *A Philosophical Essay on Probabilities* clearly demonstrates his grasp of the calculability of past events no less than it does the future, which does somewhat militate against any strict tethering of Laplace to the predictability theory. Not only does this suggest that retrodiction is as relevant a way of surmising his theory of determinism (and the activity of the intelligence in question) as is prediction. It ought also to dispel any temptation to think of Laplace as committed to pinning directionality on the world's time-evolution. This is a temptation to which, for one, Earman does to some degree succumb with his claim that Laplace in effect equates determinism and predictability[33] and, to some degree, Thomas Müller, whose asymmetric model of future-determinism is initially attributed to Laplace.[34]

Was Laplace's recognition of this symmetry coincidental, or perhaps the result of some singular insight? There is one aspect of his presentation which gives some pointers as to the provenance of this symmetry and, in so doing, prefigures modern formulations of determinism, free of the defects of the predictability criterion. Laplace's definition is explicitly mathematized: certainty about the future and past comes from the application of a formula (*formule*) to a given state of the world. Taking a formula as an emblem of determinism transports Laplace's perspective squarely into the modern debate, which is dominated by the reduction of deterministic relations between events to functions (which encode laws of nature). Laplace's commitment to the temporal symmetry of the demon's knowledge and the co-dependency of prediction and retrodiction should be no surprise given his reference to these functions. As will shortly be seen, many recent authors have taken temporal symmetry to amount to the time-reversal invariance of these functions, themselves taken to be the kernel of laws of nature. Thus, in sum, Laplace's presentation, with its reliance on predictability, suffers from the three weaknesses ((I)–(III)); but his account still respects some of the important elements of the modern view of determinism, such as the emphasis on functions. This will be a central feature of the next section.

Functional Relations

The purpose of this section is to provide the definitions of determinism and determinacy which will be relied upon throughout the rest of the book. In order to motivate these definitions, an authoritative introduction is Bertrand Russell's essay *On The Notion of Cause*, which supplies an extremely general framing of the debates over determinism and causality, with a view to trimming away several centuries' worth of what Russell deems surplus metaphysical baggage. He presents this alternative framing in terms of functional relations. These become a device for exorcising the spectre of causation from philosophical discourse, and replacing them with simpler mathematical architecture. Rather than science consisting in the discovery of regularities, science differentiates regular events into an attendant multiplicity of smaller or more complex ones. Rather than antecedent causes reproducing consequent effects, antecedents and consequents are together granulated into less coarse elements, and causes are fissile rather than reproductive. Distilled to its essence, science aims to uncover functions—more precisely, functional relations—which relate events to events. Russell proposes that no meaningful relation of cause and effect exists, but rather only formulae, cast typically in the form of differential equations, universally valid across time and space, defined for all elements of the system under consideration, which enable calculation of configuration at any instant of time in terms of the prior or later state of another. Configurations are thus related by functions rather than causes.[35] Correlatively, for Russell, scientific theories consist of formulae which relate rates of change of quantities to one another. These formulae are functions of configurations, sets of events, each of which obtains at an instant or point in time. Differential equations, for Russell, are the paradigms of these functions, whether the Euler-Lagrange equations, Newton's second law, or those which govern radioactive decay or harmonic oscillations.

Although seemingly a deflation of the notion of cause more than an intervention in the philosophy of determinism, Russell's observations can be transposed onto the terrain of the latter quite readily. This produces the groundwork for definitions both of determinism and, for the purposes of this chapter, of determinacy. Summarily, functional relationality, that is, susceptibility to functional representation, confers determinacy. In Russell's model, the world is populated by an enumerable or innumerable set of events $\{e_i\}$ occurring at times $\{t_i\}$. Each event is a configuration or state of the world, taken in its entirety—from celestial to quantum objects. Accordingly, the forces which animate nature appear in the form of functions, such as dynamical equations governing the motions of bodies which, given different times, map events at these times onto one another: for some pair $\{t_j, t_k\}$ knowledge of $\{e_j\}$ enables calculation of $\{e_k\}$ regardless which of the following hold: $j = k$, $j > k$, $j < k$.

As we will see, there are several respects in which Russell's functional relations can be weaponised for the argument of this essay. First of all, they yield an insight into what it means to speak of determinacy: insofar as they are subject to mathematical rules, any events can be mapped one onto another, given only an event at an arbitrary time. Second of all, they are a rational point of departure not only for any assessment of this conceptual datum, but the suite of related concepts this chapter aims to clarify. Finally, for the purposes of later chapters which aim to evaluate how, so defined, these concepts might square with various developments in the sciences, functional relations are couched in mathematical language which translates well across a broad range of scientific and philosophical contexts, from physics to theoretical finance and the foundations of social science, enabling these to be compared readily.

However, before exploring the implications of Russell's model in more detail and generating more specific definitions of the subject concepts, some caveats are appropriate. For several reasons, his account is far from complete and has elicited discernible opposition. For one, his steadfastly quantitative focus ignores the contribution to scientific theories made by definitions, heuristic statements, observations, or qualitative, non-functional relationships, all of which are important components of any theory. Familiar examples of the latter include Newton's first law in classical mechanics, that a force-free body persists in uniform rectilinear motion, or the laws of variation and selection proposed by Darwin in *The Origin of Species*. In this sense, scant rationale is provided for the outright autarchy of functions—other than the superficial observation that many of them appear in scientific theories, an observation which hardly warrants their inscription in the very essence of science.

These initial considerations suggest that, as a *definition* of scientific activity, or a set of necessary and sufficient conditions for a statement to qualify as a scientific law, its purview may be too limited to be practicable. It should strike no-one as surprising, then, that prominent commentators have attacked Russell's views. Nancy Cartwright rejects Russell's emphasis on these functions, for their impotence to account for functional relationships with and without explanatory power, that is, to distinguish between meaningless statistical correlations and meaningful relationships.[36] Patrick Suppes questions the adaptability of Russell's model given the emergence of non-mechanistic causal relationships, complexity, chaos and computer science.[37] Donald Lipkind presents difficulties squaring Russell's declaration of the irrelevance of the causal nexus with the ubiquity of references to the same in modern scientific discourse.[38]

These are not the only sources of controversy. In spite of his own admonition of the kinds of psychological[39] criteria in discussions of causality which might embroil them in subjectivism, Russell himself alludes to calculability as a mark of these kinds of relations between events which he takes as imperative to scientific enquiry. Adumbrating the criticisms of Laplace's formulation which is so widely deprecated today, his attack on fixes on three separate and incompatible definitions he finds in Baldwin's dictionary. Russell insists that real processes rather than one's thought, perception or experience of them must be the focus for any analysis of causality.[40] Nonetheless, the fact that configurations (sets of events, whose parameters are arguments of dynamical equations) at various times are theoretically calculable in terms of one another would seem to appeal to exactly the same subjective, inter-subjective or ideal criteria which Russell finds objectionable. Wesley Salmon argues that this vulnerability exists also in Russell's 1948 theory of causal lines, defining the possibility of inferences between events as a mark of causality.[41,42]

Thus, the extensive usage made of the framework of *On the Notion of Cause* in what follows throughout this study is no general endorsement of Russell's account, especially insofar as it might be viewed as an attempt to define scientific activity or enquiry. Nor does it presuppose it to be in any way indefeasible. Rather, it holds only that, for the purposes of Russell's attempt to show up flaws in philosophers' perspectives on causality as metaphysically bogus, these limitations are not terminal. Accordingly, *an important use can be made of the simple and broadly applicable conceptual architecture of functional relations for an analysis of temporality and determinacy*. This is the key take-away of Russell's work for the temporality of determinacy. In essence, events are primarily related functionally rather than being related causally, and functional relationships entail the determinacy of the events subject to them.

These two principles motivate the use of functions in the various research programmes discussed in this work. Having established this, it is necessary to clarify the implications of these principles for the various related terms which appear in them, determinism, causality, laws and so on.

a. Determinism

The implications of Russell's point of view for determinism are straightforward. He provides the following definition:

> A system is deterministic if, and only if, for events $e_1, e_2, \ldots, e_n$ occurring at times $t_1, t_2, \ldots, t_n$ and for any state E_t of this system at any time t, there exists a function relating the $e_1, e_2, \ldots, e_n$, $E_t = f(e_1, t_1, e_2, t_2, \ldots e_n, t_n, t)$.[43] (Def. Determinism)

Naturally, determinacy, a characteristic of the $\{e_i\}$, can be defined as subjection to some f. As one might expect, Russell limits the meaning of the participle "determining" to being related by a suitable function to other events. In this vein, determinacy can be defined in the following way:

> An event has determinacy iff it is subject to a functional relation f. (Def. Determinacy)

b. Causality

Supplying this definition of deterministic systems and relations whilst denying the meaningfulness of causal relations might lead the attentive reader to ask why these functional relations are not recognizably *causal* in any viable sense. Russell claims that events related causally display an additional characteristic, namely that the cause is asymmetrically compulsive of the effect, that is, in some way in which the effect is not compulsive of the cause. But once compulsion is evacuated of physical meaning and taken to signify mathematical determination, any lingering semblances of directionality and asymmetry disappear from the causal relation. Unsurprisingly, then, Russell's model is no more accommodating of this category than his polemical opening. The future and the past determine one-another in precisely the same sense, once determination amounts only to the subjection to functions.[44]

The avowal of temporal symmetry is stark—although not this time the symmetry of the laws encoded in Russell's functions, viz., time-reversal invariance under the mapping $t \rightarrow -t$, but rather the symmetry of relationality itself. Trivially, the state of the world at a given time is related to a successor or predecessor just as the latter is to the former. Even more stark is the equation of the *determinacy* of a variable with its being functional on others. Insofar as the relevant variables in this case are events in whose terms other events are calculable, the determinacy of an event is no more for Russell than its appearance in a well formed functional relationship, just as Quine's slogan, inherited later by Boolos, declares being to be the value of a variable.[45] According to this analysis, then, causality is consigned to the status of a relic.

c. Laws of Nature

Taking this approach to the terms "determinism" and "determinacy," the implications of Russell's account for laws of nature are predictable, too. Simply, functional relations can be recast as nomological relations. Any findable f for which $E_t = f(e_1, t_1, e_2, t_2, \ldots e_n, t_n, t)$ becomes a law of nature. Indeed, this was Russell's position, as can be observed from his straightforward equation of deterministic systems with those subject to laws; if deterministic systems are just those amenable to description by f then there are as many laws of nature as there are functions. Tying laws to quantitative, mathematical functions, Russell is in this respect the custodian of the legacy of Galileo, Kepler, Huyghens, Oldenburg and the heirs of Newtonian mechanics.

Summarily, Russell construes determinacy as functionality, determinism as subjection to functions, functions as representations of laws of nature, and such laws as time-symmetric. As it stands, however, these definitions hardly form a complete set. The onus was on this chapter to provide definitions of a number of fundamental categories (supposed to be the foundations for later chapters), to wit, determinism, determinacy, laws, laws of nature, causation and causality. Russell's minimalistic presentation of functional relations accounts naturally for most of these. But one couple remains to be defined, indeed the most problematic, as well as the most indispensible, for the questions of this work, namely temporality and atemporality. So what does the framing discussed in this section, set in terms of functional relations, have to say about this question?

Temporality and Functional Relations in Russell

Russell's reflections on the temporality of functional relations are of secondary import in *On The Notion of Cause*, mediated through his deconstruction of the metaphysics of causality and confined largely to the text's margins and digressions. They are nonetheless a fecund source of material for any analysis of the relationship of determinacy to time—a relationship which appears under the guise of the absence of the time-variable t from, or the uniformity of, the laws which confer determinacy on events. This section aims to separate the various characterisations of this relationship in the text which, as will be seen, give a good indication of the span of recent commentary on the topic. Russell makes at least four relevant assertions in the text, which can be distilled to the following tenets:

(R1)

Event e_1 occurring at time t_1 is sufficient for the event e_2 to follow—this proposition is intended to be true given any variation in t_1.[46]

(R2)

Time should not appear in functions. The uniformity of nature precludes the appearance of time as an argument of well-formed mathematical formulae (with the exception of integrated forms, for which changes in time may be considered legitimate).[47]

(R3)

The uniformity of nature is equivalent to the permanence of laws.[48]

(R4)

The uniformity of nature is not a necessary condition for the legitimacy of scientific reasoning. The permanence of all laws is less probable than the permanence of laws considered individually; the permanence for all time of an individual law is less probable than validity over a limited time. The uniformity of nature holds *a posteriori* rather than *a priori*.[49]

Take each in turn. (R1) considers conditional statements linking events to other events—statements which follow logically from laws of the form Russell prescribes—and comments on the variability or invariability of their truth values as the time value which indexes these events varies. Invariance over time underwrites the modality of the statement (Russell takes it to be necessary).[50]

Statements (R2)–(R4) all provide subtly different perspectives on the uniformity of nature. The uniformity of nature Russell, like many, takes to be an almost axiomatic principle—namely, that the world is devoid of whimsical, hypercontingent or *ex cathedra* dynamical rules; one does not, to purloin Kant's vivid imagery, wake in the night to discover that fields of yellow crops have suddenly and inexplicably turned blue, that the moon has trebled in size, or that heavy objects have become weightless, a principle which is as intuitive as it is equivocal.[51] What exactly is the meaning of these references to the uniformity of nature above and beyond registering vague antipathy to unbridled chaos—though not, of course, the kind which is now an extremely familiar aspect of dynamical systems well known to students of chaos theory?

Moving on to (R2), Russell claims that the time-variable should not appear in nomological relations, but quickly substitutes for this the claim that no laws are functions of what he refers to as absolute time (in a sense left open, but taken up at length by other authors[52]).

(R3) explicitly temporalizes the uniformity of nature, resolving it into the permanence of laws. Nature is uniform insofar as its laws are permanent. Unpacking this, though, Russell verges on tautology: he takes the example of a particular force law from which can be derived a particular acceleration, and speculates that one would expect this to hold for the future, subject to *ceteris paribus* qualifications. This amounts to the claim that every law consistent with events observed in a given time-interval holds for all time, unless it does not, in which case so does another consistent with it at earlier times, in so doing, presaging the time-dependent predicates of Nelson Goodman's new riddle of induction.

(R4) is more perplexing. It follows (R3) in equating uniformity with permanence but, at first, seems to temper its predecessor, classifying the uniformity assumption as an empirical claim and assigning a relative probability to the *prima facie* metaphysical hypothesis that all laws of nature be permanent. Russell does not detail the criteria that could be employed to identify instances of nomological impermanence—that is, to distinguish situations in which laws were discovered to fail to generalize over time from those in which laws were simply wrong, refuted by observation. Presumably, Russell holds the uniformity of nature to be in some way an empirical principle. Yet a distinction between false laws and those which are merely applicable for some strict subset of all possible times $\{t_i\}$ would appear to be a prerequisite for holding the permanence of laws to be so, whilst the latter is supposed to be equivalent to uniformity. Indeed, Earman affirms this double-bind between the plausibility and *a posteriority* of uniformity both in Russell's analysis and in other commentators'.[53]

In spite of their differences, weaknesses and ambiguities, these passages provide clear evidence that the uniformity of nature becomes a device which facilitates Russell's exploration of the nexus of temporality and determinacy. We can isolate several different theses from them, all of which pertain to this nexus.

For any law *f*, and any occurrence of any $t_l \in \{t_i\}$, the truth-value of *f* does not change with variation in t_l (Truth-Value Constancy Thesis)
For no law *f* does the time-variable *t* occur in *f* (Time-Variable Thesis)
For any law *f*, *f* is permanent (Permanence Thesis)

There are, therefore, several different meanings which must be borne in mind in order to evaluate what kind of temporality or atemporality is at work in Russell's work, and, by implication, in order to judge whether determinacy is temporal, so defined. Russell is not clear on what logical relations (if any) should obtain between permanence, the constancy of the truth-values of propositions of a certain form and permanence, other than that they all relate to uniformity in some way. One pertains to truth-values, another to the structure or syntax of certain functions, another to metaphysical characteristics. These distinctions, which at first perhaps seem trifling, are in fact crucial for focusing the efforts of future chapters, and for defining the implications of the research in physical and social science, as well as philosophy itself, for the temporality of determinacy. And whilst these more arcane of Russell's essay's pages may appear to harbour eccentricities, they in fact resonate with other prominent commentators investigating the temporality of such functions, including, as we will now see, authors of influential twentieth century texts in metaphysics and the philosophy of science.

Temporality and Functional Relations in Lindsay and Margenau

Margenau and Lindsay's classic work *The Foundations of Physics* contains a number of remarks which associate a uniformity principle, taken to be an axiom and a necessary condition for any investigation of the natural world, with the disappearance of the time-variable from functional relations and, in turn, this disappearance with effective time-independence. As is true of Russell's research, these questions are notionally linked to the philosophy of causality but, deliberately or not, feature revealing incidentals which impinge upon broader questions as to the determinacy at stake in physical laws. Their work is thus an ample repository of source material for the present question. In remarks from the closing chapter of *The Foundations of Physics*, its closest approach to standalone philosophy, the authors equate the consistency of nature with the idea that the behaviour of natural systems are absolutely independent of time, a property which they go on to associate with causality. For Lindsay and Margenau, nature can be described by differential equations, and these differential equations are free of any explicit functions of time-variables.[54] Thus, writing in the thirties, they not only repeat the themes of *On The Notion of Cause*, but begin to equate the statements which retain a degree of logical independence in the latter text. In this passage, nature's consistency (which appears to be coterminous with Russell's uniformity) is *defined* such that the relevant equations are free of the time-variable, rather than these just being two different aspects of the temporality or atemporality of laws. But whilst plainly aiming to capture an intuition as to the recalcitrance of changeable laws Lindsay and Margenau's criterion is, on the face of it, just as plainly susceptible to counter-examples.

First of all, consider the glut of laws which are not free of the time-variable in any relevant sense. In classical physics, the familiar equations which predict the movement of particles undergoing linear motion and subject to uniform acceleration ("suvat" or "davtu" equations) very clearly feature occurrences of t; their velocity relates linearly to time, as does their position vary with t^2. Could such an obvious refutation have escaped authors of their stature? This eventuality seems remote. Instead, being as these equations are kinematical equations, perhaps it can be assumed that they do not fall within the scope of Lindsay and Margenau's prescription of time-independence: perhaps the differential equations which contain no explicit functions of time need only be *higher-order* dynamical differential equations—in this case, the Euler-Lagrange equations or Newtonian laws generative of the kinds of first-order rules put to work in everyday calculations. An illustrative case here could be Newton's law of cooling: stipulating the exponential decay of temperature differences between bodies over time *prima facie* suggests time-dependency, but this disappears at the level of the differential equation from which the decay term derives, namely $\boldsymbol{J} = \boldsymbol{h}\Delta T$ (where $\boldsymbol{J}$ represents the heat flux, ΔT the temperature difference and $\boldsymbol{h}$ the coefficient of heat transfer).[55] In this way, whilst the authors do not provide an account of this distinction between rules and fundamental dynamics manifested in differential equations, a charitable reading might take this to be an implicit qualification. But does this response remove the difficulties, not only with the time-dependency of lower-order kinematical rules, but also higher-order laws?

To explore this, consider a simple example. In the context of the thermal physics, Boltzmann's transport equation concerns the time-evolution of the probability density function $f(\boldsymbol{r}, \boldsymbol{v}, t)$, a function which specifies the statistical distribution of positions and speeds (or momenta $\boldsymbol{p}$) in a gas.[56,57] The transport equation expresses the total derivative over time $\frac{df}{dt}$ as a sum of several terms which correspond to the temporal effects on the gas of diffusion, molecular collisions and the action of external forces. Very clearly, concerning as it does the evolution of a probabilistic function, this law of Boltzmann's manifests a temporal dependency in the rather brute sense that the time-variable features in the equation, without any obvious avenue to eliminate it. Insofar as it pertains to rates of change of physical quantities, just as do a profusion of other similar rules, time-dependency would appear to be inscribed into the very essence of science, but hardly in a way which compromises the uniformity of nature—an indication that the absence of the former fails as a credible necessary condition for the latter. Clearly there is a distinction between the occurrence of the time-variable in the equation being scrutinised and the relevant sense of time-dependency or time-independence of the system in question, being, for the authors, uniformity. As much as anything, these considerations do highlight the high degree of arbitrariness involved in couching metaphysical hypotheses loosely in terms of the appearance or non-appearance of particular variables which, after all, are *representations* of systems: by substitution, one can always replace occurrences of the variable t with another and give it a different description, but this does not secure the condition that the equation is independent of time in the requisite sense.

Do these extremely basic counter-examples mean that Lindsay and Margenau's account fails as an accurate lens for the investigation of the temporality of determinacy, or as a serviceable criterion for the uniformity of nature, defined in terms of the occurrence or non-occurrence of t in dynamical equations? It is easy to see from their writings their likely riposte. First of all, whilst rudimentary kinematical equations, Newton's laws, exponential functions governing radioactive decay, and a multitude of other quantities clearly feature occurrences of time-variables—sometimes occurrences which themselves seem to change over time in some way—many of these occurrences are superficial and do not pertain to the dynamical fundamentals of the theories in question. The authors qualify that it is the *differential equations* describing nature[58] which do not contain functions of time—not equations *simpliciter*. What, then, is the true scope of their prohibition on functions of time, and does this successfully fortify their case against counter-examples? The standalone writings of Margenau as well as the commentary on this work by A. P. Ushenko in his papers "The Problem of Causal Inference" and "The Principles of Causality" offer useful elaboration. Modifying Lindsay and Margenau's criteria, Ushenko proposes that causal laws be regarded as second-order differential equations, with no explicit appearance of the time-variable t.[59] He seems to have in mind here something resembling the following schema for causal laws:

$$df'' + bf' + cf + d = g\,(\mathrm{F1})$$

where $f' = \frac{\partial f}{\partial t}$, $f'' = \frac{\partial^2 f}{\partial t^2}$.

In this schema, Ushenko (vicariously Lindsay and Margenau) would have it that t not appear in g (regardless of whether it appears in f for instance) in order to qualify as a causal law (and therefore confer uniformity and, ultimately, atemporality). Newton's second law, for instance, mimics such a schema. Thus, the occurrence of the time-variable in rates of change in formulae does not refute Margenau and Lindsay's prescription. The evolution of the system over time *means* first and second order variations in the relevant functions over time; the question as to the temporal dependency of the laws of nature is decided on whether the *further* quantities on which these variations depend feature the occurrence of t.

These supplementations to Lindsay and Margenau's *Foundations of Physics*, *The Nature of Physical Reality* and other texts in the philosophy of science do indeed inoculate their approach against simplistic counter-examples such as were sketched above, velocities linearly scaling over time with a constant acceleration, being kinematical rules and calculations rather than laws in the strict sense. But it is difficult to see how it dispels the totality of possible such examples, particularly those which resemble straw men less. In the context of general relativistic cosmology, the cosmological scale factor $a(t)$ used to parameterise the universe's spatial expansion, featuring in the Friedmann differential equations and the Hubble parameter, boasts a very clear temporal dependency. Indeed, this scale factor not only features occurrences of t; these occurrences themselves involve different powers of t depending on the cosmological era—$t^{1/2}$ or $t^{2/3}$ respectively for the radiation-dominated and matter-dominated phases of expansion:[60]

$$\left(\frac{a}{a_{eq}}\right) = \begin{cases} \left(3/2\sqrt{2}\right)^{\frac{2}{3}} \left(H_{eq}t\right)^{\frac{2}{3}} \\ \left(3/\sqrt{2}\right)^{\frac{1}{2}} \left(H_{eq}t\right)^{\frac{1}{2}} \end{cases} \quad (\mathrm{F2})$$

In particular, the two differential equations of which a is a solution

$$\frac{\dot{a}^2+k}{a^2}=\frac{8\pi G}{3}\rho(t)(\mathrm{F3})$$

$$\frac{2\ddot{a}}{a}+\frac{\dot{a}^2+k}{a^2}=-8\pi Gp(t)(\mathrm{F4})$$

do not seem amenable to the schema (F1) above.[61] The physicist Paul Dirac, in a similar vein, considers that the discoveries of cosmology can be accounted for only by relativizing time to the epoch, with the possibility of different solutions, different dynamics, and altogether different laws from epoch to epoch.[62] As will be discussed later on, these accounts rarely scrutinise the relationship between the era and t, that is, the time of the system being studied and the time in which these systems themselves exist, and in which transitions from system to system occur. The bifurcation of the ratio of a to a_{eq} in (F2) into two equations in t concretizes the temporal dependency of the model itself, and the dependency of the scale factor on two separate orders of time. But, regardless of this fact, the inaptitude of (F1) in the context of (F2)–(F4) is inexplicable by the criterion of Margenau, Lindsay and Ushenko, even on the most charitable reasonable reading. Indeed, in a footnote, Ushenko begrudgingly acknowledges the limited scope of his ostensibly general declaration.[63] For his part, Margenau relativizes his insistence that uniformity entails time-independence with a caveat around causal closure. Even if equations could be found which violate their criteria for uniformity and consistency, this could be down to the failure to completely understand the phenomena being described and the coarseness of the analysis involved or, alternatively, because an extraneous causal influence not fully described by these equations perturbs the system—were the latter fully represented and the system causally closed rather than open, no such discrepancies would arise.[64] According to this response, then, uniformity is a sufficient condition for time-independence, in the sense that the relevant equations are free of t; appearances of t signify the causal openness of the system, and only causally closed systems can be uniform.

How successful is this defence? Does this insulate Lindsay, Margenau and Ushenko from these counter-examples, and legitimate the conditions they attempt to impose on appropriately formed dynamical laws? The non-circularity of their overall presentation clearly depends on successfully unpacking the concepts of causal closure and openness in a way which does not just collapse back into the uniformity of nature. It is dubious as to whether the text meets this condition. The authors' case is reliant on the thesis that appearances of the time-variable in the relevant dynamical equations camouflage external forces, violating causal closure, and on the possible swapping out of this time-variable for a suitable description of these forces. Even bracketing the question as to the general validity of this approach, it squares uncomfortably with the previously mentioned case of cosmological expansion—how is causal closure of the sort envisaged by Margenau here to be brought to bear on this area of research, and what of the discontinuities posed by the bifurcating scale factor, that is, the fission of the function $a(t)$ into different components valid for different epochs? The authors do not provide a demonstration. It could indeed prove altogether impossible to extrapolate the criteria of Lindsay and Margenau to the cosmological scale, in which case the scope of their analysis would have to be attenuated accordingly.[65]

Pulling all of this together, how does the work of Lindsay and Margenau (and their commentators) impinge upon the original question—the temporality involved in functional relations? To answer this, recall the distinctions implicit in Bertrand Russell's account, the distinction of several different formulations of this relationship.

For no law f does the time-variable t occur in f (Time-Variable Thesis)

The work of Lindsay and Margenau allows this Time-Variable Thesis to be clarified and refined. For them, first of all, laws are differential equations, and the scope of the prohibition on occurrences of t is accordingly limited to these differential equations. Secondly, this prohibition does not debar functions which describe the *time-evolution* of systems, expressed typically in the form of partial or total derivatives; rather, it debars occurrences of t from the quantities on which this evolution itself depends:

$$df'' + bf' + cf + d = g \left(\text{F1}\right)$$

where $f' = \frac{\partial f}{\partial t}$, $f'' = \frac{\partial^2 f}{\partial t^2}$.

Thirdly, their definition applies only to causally closed systems—a restriction which, as has been noted, does betray a certain circularity.

These aspects of Lindsay and Margenau's writings, then, allow the Time-Variable Thesis to be defined. They are therefore a fertile source for developing the meaning of temporality and determinacy originally found in Russell. However, there is another noteworthy aspect of their writings which equally well expresses their atemporal approach to laws, and is well worth examining. At times, their conception of the temporality of functional relations, normally the presence or absence of t, is recast in terms of the *invariance* of *time translation*. In another excerpt from the *Foundations*, they claim that the consistency of nature requires not only the absence of any absolute time-dependency, but also something more specific: that solutions of the equations describing the system which are boosted by a constant distance in time remain solutions afterwards. Substituting $x = f(t - t_0)$ for $x = f(t)$, in other words, translates solutions of the relevant equations to solutions. As an example, they consider the exponent in Coulomb's force law: it should have a particular value regardless of where in time one chooses to study it, and should remain so forever.[66]

Thus, Lindsay and Margenau's criterion has to do with the preservation of solutions of equations under temporal mappings: the substitution of $t - t_0$ for t in f preserves solutions of equations which are laws of nature (or, equivalently, the mapping $t \rightarrow t - t_0$ takes solutions to solutions). This is a distinctive contribution of theirs, somewhat separate to Russell's, which suggests the need to distinguish and examine the following further interpretation of the temporality of functions:

Under the mapping $t \rightarrow t - t_0$ solutions of dynamical equations map to solutions of dynamical equations. (Translation Invariance Thesis)[67]

How convincing is this alternative view which Lindsay and Margenau put forward? One concern is that it is obscure what connection this relationship bears to the *definiendum*—the consistency or uniformity of nature as a proxy for the time-independence or atemporality of functional relationships. Does it hold by inspection that time translation invariance entails that *f* is independent of *t*, as well as the converse? By way of a simple illustration, returning to the cosmological context, with the distinction between radiation-dominated and matter-dominated phases of expansion, we can straightforwardly envisage times t_p, t_q where, for t_{eq}, the time at which the radiation and matter have equal densities, $t_p < t_{eq}$, $t_q > t_{eq}$. Thus, under the mapping $t_p \rightarrow t_q$, different solutions of (F3) and (F4) in *a* appear. Of course, the authors could always disqualify such instances on the basis that the requisite causal closure conditions fail for those generally relativistic circumstances in which, in Earman's view, no tenable distinction can be upheld between geometrical structures of space-time and their physical contents, and absolute and dynamical objects.[68] But as increasing numbers of *ad hoc* stipulations are required to sustain the Translation Invariance Thesis in the face of a tapestry of counter-examples, the closer this putatively empirical principle approaches the terrain of tautology, analyticity or *a priority*. This is an obvious but important weakness in the Translation Invariance Thesis.

Despite its obvious weaknesses, the translation invariance thesis is not some idiosyncratic non-sequitur of these authors, but rather an equivalence which is accepted quite routinely in the literature. Feigl and Brodbeck's reflections on "Causality, Determinism and Probability" in their compendium *Readings in the Philosophy of Science* provide an example. Once again presenting constraints on the temporal structure of functions as necessary conditions for the consistency of nature, Feigl repeats the contention of Lindsay, Margenau et al. as to the occurrence of the time-variable in nomological functions: spatio-temporal positions, in themselves, do not have any dynamic influence over real events. Space and time (or the rather variables representing them) do not explicitly occur in the functions which represent natural laws, just as spatial directions might be taken as irrelevant to the way in which forces propagate. Difference in the effects brought about by natural laws must be accounted for by differences in initial conditions, and not by differences in spatial or temporal locations.[69] Feigl regards these aspects of time-independence as expressions of the homogeneity of time, a principle he attributes to Maxwell's *Matter and Motion*:

Spatio-temporal positions, in themselves, do not have any dynamic influence over real events.[70] (Homogeneity Thesis)

Like Lindsay and Margenau, he takes a close logical relationship, perhaps one of outright equivalence, to obtain between homogeneity, translation invariance, and the explicit appearance of time in the relevant functions. Events' positions in space and time do not effect or affect them. This, he holds, is equivalent to the functions describing these events being explicitly free of the space and time variables. Homogeneity and translation invariance are then taken to be jointly necessary conditions for his causal principle (same causes, same effects). Thus, the uniformity of nature (in the guise of this supposedly axiomatic causal principle) entails the homogeneity of time and translation invariance, and these entail in turn the disappearance of time from the relevant equations: Uniformity of nature → Homogeneity Thesis, Translation Invariance Thesis ↔ Time Variable Thesis.

It is hard not to conclude that these various logical connections are asserted bluntly and without the requisite justificatory work. Even if the homogeneity of time requires that equations be free of t, in what sense is this equivalent to the truth or falsity of translation invariance, and why does causality entail the latter?[71] It is hard to see why the fact that temporal dependencies are explicit rather than implicit should influence metaphysical questions as to determinism and causality, especially in view of the vast descriptive arbitrariness at stake in any equation of motion (such as is implied by the possibility of relabeling variables), even when one introduces considerations of simplicity of form. Similar lacunae present themselves as have been mentioned in the context of Lindsay and Margenau—the mysterious logical connection between the time at which events occur and the occurrence of certain variables in nomological functions. Whereas the former would appear to be a statement about the explanatory redundancy of absolute time or, perhaps, a symmetry such as translation invariance of the parameter in question, the latter concerns the *form* equations take. And, as already observed in the context of Earman and Russell's analyses, the statement that differences in effects imply differences in conditions embeds rather limited empirical content, threatening to tautologise Feigl's perspective on determinism and causality.[72,73] What these passages do demonstrate, however, is that Feigl, like Lindsay, Margenau and others, upholds the Translation Invariance Thesis as an expression of the atemporality conjecture as well as the logical relationship between this and the time-independence of laws, ensuring the logical arrow from supposedly axiomatic causal or deterministic principles to the atemporality of determinacy.

Conclusion

What, then, does this survey of the literature imply as to the problem in hand, namely the temporality of functional relations? More fully put: for those authors who take determinacy to be conferred by the subjection of events to laws, and those laws to be encoded by mathematical functions, what notion of temporality is applicable to them? Answering this involves drawing on all these contributions to the debate from Russell, Feigl, Earman, Ushenko, Lindsay, Margenau and more—not to mention the other recent scholars who address related questions of truth-values, laws and time.[74] We can therefore assemble a number of sub-doctrines which amount to an *atemporality conjecture*:

Under the mapping $t \rightarrow t - t_0$ solutions of dynamical equations map to solutions of dynamical equations. (Translation Invariance Thesis)

For any law f, and any occurrence of any $t_l \in \{t_i\}$, the truth-value of f does not change with variation in t_l (Truth-Value Constancy Thesis)

For any law f, f is permanent (Permanence Thesis)

For no law f does the time-variable t occur in f where $dg'' + bg' + cg + d = f$ and $g' = \frac{\partial g}{\partial t}$, $g'' = \frac{\partial^2 g}{\partial t^2}$. (Time-Variable Thesis)

These, then, are the key expressions of the atemporality conjecture. From the Stoics to the scholastics, and from theological determinists to philosophers of physics, the intuition that the determinacy of events, their subjection to rules, be *atemporal* has been variously recapitulated, cloaked either in terms of the eternity of divine or infernal sapience, or the nomological status of functional relations. In these accounts, the atemporality conjecture asserts the time-independence of the laws of nature.[75] These theses provide more precise stipulations of this conjecture. The logic at work in these can be summarised in the following argument for the time-independence of the laws of nature:

(P1) Laws of nature are functional relationships.
(P2) Functional relationships confer determinacy on events.
(P3) The determinacy conferred on events by functional relationships is atemporal.
(P4) Conferring determinacy on events atemporally implies time-independence.
(C) Laws of nature are time-independent.

(P3) is a central expression of the atemporality conjecture. The sense of the term "atemporal" encompasses all the competing forms of this conjecture listed above, from the Time-Variable Thesis to Translation Invariance. These distinctions, as we have seen to be elided frequently in many parts of the academic literature, will be indispensable for focusing the discussion of the remaining chapters of this work, which attempt to catechize this atemporality conjecture and to test it against research and developments in the physical and non-physical sciences. Ultimately, this essay will find that the conjecture will prove impossible to reconcile with them. The Truth-Value Constancy thesis fails for systems with recursive temporal dynamics, such as markets for equities and derivative securities, in which participants' expectations about others' expectations generate price volatility, or for ergodic systems, which display probabilistic irrelevance or independence. The Time-Variable Thesis in all its guises sits uncomfortably with models in time series analysis, predictive modelling or economic forecasting, which contain numerous permutations of the time-variable. The Permanence Thesis becomes superfluous when faced with the possibility of uniform and consistent laws which are nonetheless impermanent and time-dependent. These and other objections mobilised against the atemporality conjecture throughout the following chapters will build a case for it to be rejected. In its place, they attempt to articulate an alternative vision of the temporality of determinacy.

Notes

1. Plato (2002), p. 70.
2. Earman (1986), p. 4.
3. Earman (2004a), p. 1.
4. Earman (1986), p. 233.
5. Mellor (1995), p. 200.
6. Hoefer (2016).
7. Woodward (2019).
8. Kant (1985), p. 4.
9. Gilles Deleuze and Nancy Cartwright are examples of philosophers who deny that the discovery of laws, considered as universal regularities, is a fundamental aspect of scientific enquiry.
10. Carroll (2016).
11. Zilsel (2003), p. 120.
12. Zilsel (2003), p. 120.
13. Russell (1918).
14. Earman (1986), p. 11.
15. Miller (1971), p. 73.
16. Laplace (1902), p. 4.
17. The passages following Laplace's endlessly cited extract are even stronger in their epistemic inclination, referring to what can be grasped by the human mind, intelligence, and the knowledge and discoveries which result from it. Laplace (1902), p. 4.
18. Hoefer (2016).
19. Primas (2002), p. 90.
20. Van Kampen (1991), pp. 273–274.
21. Popper (1982), pp. 1–2, 11.
22. One point of divergence must be noted. Attentive to the tendency for experimental results which are unexpected (whether unpredictable or contrary to predictions) to be explained away as insufficiencies in such descriptions, and to the concomitant risk that his definition ends up qualifying meaningless theories as deterministic, Popper does depart

from Laplace by making the relationship between the precision level of these descriptions and the precision level of the predictions an object of the scientific theory under consideration.

23. Werndl (2016), p. 2.
24. Werndl (2009a).
25. Norton (2003), p. 9.
26. Werndl (2009a).
27. Earman (1986), p. 9.
28. Primas (2002), p. 90.
29. Werndl (2016), p. 1.
30. Hoefer (2016).
31. Roughly, the mapping $t \rightarrow -t$ translates sets of solutions of the equations which encode these laws to sets of solutions, *pace* Earman's 1974 paper, in which he shows that this convenient framing of time-reversal invariance can become intractable in theories which admit relativistic space-times, specifically for space-times which possess no global function relative to which it displays temporal isotropy. This, Earman claims, inhibits construction of mirror image states in time, of which one is the negative of the other, due to topological and geometrical differences in the relevant hypersurfaces. This sits uncomfortably with the standard formulation of time-reversal invariance. See Earman (1974), p. 30.
32. Hoefer (2016).
33. Earman (1986), p. 7.
34. Müller (2015), p. 735. A model $\langle W, T, <, V \rangle$ counts as future-deterministic if and only if given any two ω, $\omega' \in W$, if there is a time $t \in T$ for which $V(\omega, t) = V(\omega', t)$, then for any other time $t^* > t$, it is also the case that $V(\omega, t^*) = V(\omega', t^*)$.
35. Russell (1918), p. 190.
36. Cartwright (1983), pp. 21–43.
37. Suppes (1970).
38. Lipkind (1979), p. 702.
39. Russell (1918), p. 184.
40. Russell (1918), p. 184.
41. Russell (1948), p. 459.
42. Salmon (1984), pp. 140–145.
43. Russell (1918), p. 199.
44. Russell (1918), p. 195.
45. Quine (1961), p. 15.

46. Russell (1918), p. 183.
47. Russell (1918), p. 205.
48. Russell (1918), p. 196.
49. Russell (1918), pp. 196–197.
50. Logicians such as Kazimierz Trzęsicki in his "Indeterministic Temporal Logic" also appear to draw this connection between determinism and the constancy of truth values of propositions describing events. See Trzęsicki (2015), p. 140 for salient definitions of determinism, pre-determinism, and post-determinism in these terms.
51. In spite of the intuitive appeal of Russell's dictum, even in this form, has been subject to scrutiny, as we will see more fully in future chapters. For a succinct example, see Smolin (2013), pp. 114–122 who follows C. S. Peirce's elliptical challenge to the uniformity of nature in view of a history of these laws (see Peirce (1891)).
52. See e.g.s Ushenko (1942), Lindsay and Margenau (1963), Feigl and Brodbeck (1953).
53. Earman (1986), p. 134.
54. Lindsay and Margenau (1963), p. 522.
55. Blundell and Blundell (2010), pp. 99–100.
56. Blundell and Blundell (2010), p. 31.
57. Brown and Myrvold (2008), pp. 2–3.
58. Lindsay and Margenau (1963), p. 522.
59. Ushenko (1942), p. 136.
60. Padmanabhan (1993), p. 64.
61. Padmanabhan (1993), p. 60.
62. Dirac (1939).
63. Ushenko (1942), p. 136.
64. Lindsay and Margenau (1963), p. 523.
65. Lee Smolin describes this attenuation as akin to conducting physics within a box, taking it to be a deep-rooted and general defect proper to contemporary physics (from Newtonian mechanics to special relativity) with an essential relationship to the atemporality of laws. See Smolin (2013), p. 94.
66. Lindsay and Margenau (1963), p. 522.
67. Earman provides a more sophisticated definition of this property which rests on a notion of possible worlds ω_L (a commitment it is convenient to circumvent for the purposes of this chapter) which are consistent with putative laws L: Laws L are *time-translation invariant* if, and only if, for any real Δ, assuming $\omega \in \omega_L$, then $\omega^\Delta \in \omega_L$, where $\omega^\Delta(t)$ is defined to be

equivalent to another function $\omega(t + \Delta)$, for any and every time t. See Earman (1986), p. 129.
68. Earman (1986), p. 135.
69. Feigl and Brodbeck (1953), p. 412.
70. Feigl and Brodbeck (1953), p. 412.
71. There are well known principles in theoretical physics which relate translation invariance to conservation principles, such as Noether's theorem. This might seem to provide some rationale for the various connections being drawn by Feigl and others. If the denial of time translation invariance would entail the failure of conservation in these cases then taking conservation to be a fundamental principle would force time translation invariance to be affirmed, assuming Feigl's equivalence relation between this and the Time Variable thesis, grounding this form of the atemporality conjecture. But it is far from obvious that this relation (given in e.g. Noether's first theorem) possesses the metaphysical generality needed to complete the argument: As Earman emphasises, such a move requires that the laws be Euler-Lagrange differential equations derivable from an action principle, that the symmetries involved be continuous (and, specifically, form a certain Lie group, and be variational symmetries. See Earman (2004), pp. 1231–1232. The qualifiers which Earman is careful to insert are not always respected in discussions of the theorem, as will be discussed more fully in subsequent chapters of this work.
72. Earman (1986), p. 134.
73. Russell (1918).
74. For a prominent example of the discussion of sentential truth-values and time see Richard (1981). Richard distinguishes between eternalist and temporalist perspectives on whether sentences express propositions whose truth-values can change over time, and argues that the latter of the two fails to account for belief attribution.
75. Lindsay and Margenau (1963), p. 522.

Bibliography

Blundell, K., & Blundell, S. (2010). *Concepts in Thermal Physics*. Oxford University Press.

Brown, H., & Myrvold, W. (2008). Boltzmann's H-theorem, Its Limitations, and the Birth of (Fully) Statistical Mechanics. Retrieved May 23, 2020, from https://arxiv.org/abs/0809.1304

Carroll, J. (2016). Laws of Nature. *The Stanford Encyclopedia of Philosophy*. Retrieved May 3, 2020, from https://plato.stanford.edu/archives/fall2016/entries/laws-of-nature/

Cartwright, N. (1983). Causal Laws and Effective Strategies. In *How The Laws of Physics Lie* (pp. 21–43). Oxford University Press.

Dirac, P. (1939). The Relation Between Mathematics and Physics. *Proceedings of the Royal Society (Edinburgh), 59*, 122–129.

Earman, J. (1974). An Attempt to Add a Little Direction to "The Problem of the Direction of Time". *Philosophy of Science, 41*(1), 15–47.

Earman, J. (1986). *A Primer on Determinism*. D. Reidel.

Earman, J. (2004a). Laws, Symmetry and Symmetry Breaking: Invariance, Conservation Principles, and Objectivity. *Philosophy of Science, 71*(5), 1227–1241.

Feigl, H., & Brodbeck, M. (Eds.). (1953). *Readings in the Philosophy of Science*. Appleton-Century-Crofts.

Hoefer, C. (2016). Causal Determinism. In E. Zalta (Ed.), *The Stanford Encyclopedia of Philosophy*. Retrieved May 2, 2020, from https://plato.stanford.edu/archives/spr2016/entries/determinism-causal/

Kant, I. (1985). Metaphysical Foundations of Natural Science. In J. Ellington (Trans.), *Philosophy of Material Nature*. Hackett.

Laplace, P. (1902). *A Philosophical Essay on Probabilities* (F. Truscott & F. Emory, Trans.). John Wiley & Sons.

Lindsay, R., & Margenau, H. (1963). *Foundations of Physics*. Dover Publications.

Lipkind, D. (1979). Russell on the Notion of Cause. *The Canadian Journal of Philosophy, 9*(4), 701–720.

Mellor, D. (1995). *The Facts of Causation*. Routledge.

Miller, J. (1971). The Principle of Causality: Tautology, Empirical Proposition or Imperative? *The Southwestern Journal of Philosophy., 2*(1/2), 73–82.

Müller, T. (2015). Time and Determinism. *Journal of Philosophical Logic, 44*(6), 729–740.

Norton, J. (2003). Causation as Folk Science. *Philosopher's Imprint, 3*, 4. Retrieved May 10, 2020, from https://quod.lib.umich.edu/p/phimp/3521354.0003.004/1

Padmanabhan, T. (1993). *Structure Formation in the Universe*. Cambridge University Press.

Peirce, C. (1891). The Architecture of Theories. *The Monist, 1*(2), 161–176.

Plato. (2002). Meno. In G. Grube (Trans.), *Plato: Five Dialogues*. Hackett.

Popper, K. (1982). *The Open Universe: An Argument for Indeterminism.* Routledge.
Primas, H. (2002). Hidden Determinism, Probability and Time's Arrow. In H. Atmanspacher & R. Bishop (Eds.), *Between Chance and Choice. Interdisciplinary Perspectives on Determinism* (pp. 89–113). Imprint Academic.
Quine, W. (1961). On What There Is. In *From a Logical Point of View* (2nd ed., pp. 1–19). Harper and Row.
Richard, M. (1981). Temporalism and Eternalism. *Philosophical Studies: An International Journal for Philosophy in the Analytic Tradition, 39*(1), 1–13.
Russell, B. (1918). *Mysticism and Logic and Other Essays.* Longmans, Green and Co.
Russell, B. (1948). *Human Knowledge.* Simon and Schuster.
Salmon, W. (1984). *Scientific Explanation and the Causal Structure of the World.* Princeton University Press.
Smolin, L. (2013). *Time Reborn.* Penguin.
Suppes, P. (1970). *A Probabilistic Theory of Causality.* North Holland.
Trzęsicki, K. (2015). Indeterministic Temporal Logic. *Studies In Logic, Grammar and Rhetoric, 4*(1), 139–162.
Ushenko, A. (1942). The Problem of Causal Inference. *Philosophy of Science, 9*(2), 132–138.
Van Kampen, N. (1991). Determinism and Predictability. *Synthese, 89*(2), 273–281.
Werndl, C. (2009a). What Are the New Implications of Chaos for Unpredictability? *The British Journal for the Philosophy of Science, 60*(1), 195–220.
Werndl, C. (2016). Determinism. In M. Griffith, T. Neil, & K. Timpe (Eds.), *Routledge Companion to Free Will.* Routledge. Retrieved May 10, 2020, from http://philsci-archive.pitt.edu/12166/1/DeterminismIndeterminismWordPittsburghArchiveWithF.pdf
Woodward, J. (2019). Scientific Explanation. In E. Zalta (Ed.), *The Stanford Encyclopedia of Philosophy.* Retrieved May 2, 2020, from https://plato.stanford.edu/archives/win2019/entries/scientific-explanation/
Zilsel, E. (2003). The Genesis of the Concept of Physical Law. In *The Social Origins of Modern Science* (pp. 96–122). Kluwer Academic Publishers.

3

The Temporality of Determinacy I: Philosophy of Non-Physical Sciences

Introduction

The research of the sociologist Andrew Abbott disinters a number of covertly metaphysical premises at work at the foundations of the social sciences. Under the pretext of aligning its methodology with the physical sciences, purporting to model and explain the behaviour of the social world, or capturing relations between variables generated by underlying causal processes, empirical sociology ends up trafficking into the discipline a basket of questionable assumptions. Abbott gathers these under the label GLR or General Linear Reality.[1] Despite its effectiveness, he holds it to have distorted scholars' construal of social reality, and limited their thinking in a way which can be corrected only by embracing a broader suite of formal tools,[2] a regrettable side-effect of the model (not to mention the philosophical infractions involved). In his "Transcending General Linear Reality" he highlights six assumptions at the core of the GLR doctrine, some of which pertain to the metaphysics of individuation, others to the relationship of theoretical descriptions to reality. These he outlines in the abstract, locates in specific sub-disciplines—whether demographic trends, shifting class antagonisms, or the econometrics of the labour market—and then studies critically, highlighting their deficiencies both as sociological tools and ontological[3] generalisations.

C. Husbands, *The Temporality of Determinacy*,
https://doi.org/10.1007/978-3-030-86530-6_3

Time has a privileged status in Abbott's work. In this sense, his programme is a useful lens for the specifically *temporal* conceptions at work in the core of the social sciences, conceptions which a philosopher of time might consider controversial. Abbott thus makes it his objective to reinject time for the better not only into statistical assumptions but also into philosophical ones,[4] and into the laws which determine the entities populating social reality. He attacks the atemporality (not to mention the aspatiality) of sociology when brought to bear on the issue of socio-economic inequality in his essay "Inequality as Process."[5] He considers the possibility that distant time-points can be causally efficacious on others and speaks of the need to acknowledge action at a temporal distance[6] rather than contiguity between related events. Elsewhere, in his compendium *Processual Sociology*, he allegorizes the social world as a monster of change undergoing repeated deaths and rebirths, and its constituent parts, whether individuals, cultures, institutions or their interactions, as processes beginning, repeating and unfolding.[7]

It is at first interesting to note, however, that many of these seeming forays off the sociological reservation and into philosophy are actually perfectly consistent with traditional, atemporal approaches to the determinacy of social systems. Despite the *prima facie* militancy of this position, which conjures images reminiscent of Nietzsche's metamorphosing cosmos[8] expatiated in the Nachlass and Zeitatomenlehre, there is nothing yet specifically *temporal* about Abbott's intervention into the relations sociology concerns. The contention that things are made and remade (which can be chalked up to the statement that events happen), that this process occurs from instant to instant, and that this transition from instant to instant is continuous, signifies little above and beyond the mundane hypothesis that events are parameterised by time, and that the latter is a continuous rather than discrete variable. The participle "unfolding" confers no more radical a vision of temporality than is implied by A-serial or tensed time. Indeed, the counter-intuitive notion of historical action at a distance (the possibility of temporally non-contiguous or non-adjacent events influencing one another) becomes a truism once the notion of influence between events is suitably deflated into functional relationality, as in the previous chapter. Events arbitrarily far apart (and irrespective of their order) would influence each-other in the trivial sense that they are functionally related, and that a determination of one can be made given a specification of the other.

So, in spite of Abbott's deprecation of the atemporality of the typical sociological perspective, the above extracts are, taken in themselves, largely reconcilable with many of the formulations of the conjecture deconstructed in the previous chapter. The permanence of laws is hardly debarred by the impermanence of events, nor is the invariance of their truth-value with respect to time. Nor does, *a priori*, the absence of the time-variable from higher order dynamical equations estop events occurring "in time." Where, then, is the more radical kernel at stake in Abbott's critique of the role of temporality in social sciences, which inspires his generalized cynicism about his discipline's benighted state? Identifying this allows us to see the way in which his work aligns with the temporality of determinacy.

In *Time Matters* Abbott provides several clues, including one which is particularly apposite given the argument made in this chapter. The general-form GLM frames society as a set of entities (usually individuals) with attributes, and accounts for their behaviour as a series of transformations of prior states: $\boldsymbol{X}_t = \boldsymbol{X}_{t-1}\boldsymbol{B} + \boldsymbol{U}$.[9] Abbott identifies a number of assumptions at the heart of this form of the GLM which pertain to the temporality of determinacy, and accordingly to our present focus. Two demonstrative examples of these are the following:

> Causal or temporal orders are not defined when attributes of entities being described vary between different time periods.[10]
>
> The order in which events occur does not in itself influence outcomes. For the general-form GLM the state of a given entity x_i at a time $t + 1$ is the outcome of applying the transformation matrix $\boldsymbol{B}$ to its prior state at another time t.[11]

These statements capture the GLM's exclusion of sequence effects. For Abbott, understanding the structure of important relationships, such as the relationship between criminality and unemployment, requires distinguishing the order in which the events occur. For example, an individual who is criminalised after becoming unemployed is a rather different data point compared with the reverse. The GLM model, however, accommodates no such distinction: any given event is the result of the application of a transformation matrix to its predecessor, such a matrix being independent of time, and bearing no relation to whether and how previous events have occurred. To take adequate account of Abbott's insistence on the pertinence of order, the model would have to be such as to permit sequential dependencies such that for some pair of times t_i, t_j which feature in $\boldsymbol{B}$, $\boldsymbol{B}^* \neq \boldsymbol{B}$, where $\boldsymbol{B}^*$ is defined as $\boldsymbol{B}$ where $i > j$ and as the transformation of $\boldsymbol{B}$ under the mapping $\boldsymbol{X}_i \leftrightarrow \boldsymbol{X}_j$ otherwise. As one might expect, Russell's broad functional model $f(e_1, t_1, e_2, t_2, \ldots e_n, t_n, t)$ has accordingly little to say of such translations. No obvious aspects of the function f discriminate between pairs of events ordered in one way over against another, nor suggest anything as to the behaviour of this function under the mapping $m \leftrightarrow n$, for some pair of events $\{e_m, e_n\}$. By contrast, Abbott's analysis of the unemployment-criminality nexus suggests clearly that for t_m, some time at which an individual becomes unemployed, and t_n, some time at which an individual commits a crime, the relevant function or matrix is not invariant under this mapping. This insight undermines an important hypothesis which is characteristic of the GLM—namely that the causal role of particular attributes are independent of spatio-temporal context. Abbott claims that the formal equivalent of this hypothesis is the thesis that the coefficients in the matrix $\boldsymbol{B}$ in the general GLM are independent of $\boldsymbol{X}_t$, $\boldsymbol{X}_{t-1}$, $\boldsymbol{X}_{t-2}$, etc.[12]

Abbott inveighs against this hypothesis on the basis that systems differing only in their temporal context and not in any of the $\boldsymbol{X}_i$ follow divergent evolutions and therefore possess different $\boldsymbol{X}_t$ for larger (more futural) values of t. By way of illustration of this seemingly abstract framing, he takes a simple example, considering Bradshaw's account of two African nations' agricultural policies in international development theory. In this account, several key variables purport to explain development (for instance, degree of product specialization). In spite of this key variable in the trend model being identical across different cases, very different development trajectories result—for Abbott, to be chalked up to differences in temporal context which cannot be expressed as a superposition of interaction effects. If the causal role of an attribute is taken as its role as a determinative or determining influence on another event, it can be equated with an event conceived in its *functional* relationship to other events. If what Abbott refers to as an attribute's temporal context is its temporal location, then independence from this context entails independence from temporal location. And, on the assumption that independence from temporal location entails time-translation invariance, comparisons can easily be made between this aspect of the GLM and two formulations of the atemporality conjecture:

> Under the mapping $t \rightarrow t - t_0$ solutions of dynamical equations map to solutions of dynamical equations. (Translation Invariance Thesis)
>
> For any law f, and any occurrence of any $t_l \in \{t_i\}$, the truth-value of f does not change with variation in t_l (Truth-Value Constancy Thesis)

Thus, a line of attack already seems to have been opened up through the prism of Abbott's analysis of the GLM.

Many more examples of objectionable atemporality theses surface throughout Abbott's work, and are deconstructed and attacked in the contexts of a variety of sub-disciplines.[13] Blazoned consistently and courageously throughout his academic career, this approach strikes out at what he takes to be problematic yet uninterrogated methodological assumptions. But today, despite his seeming contrarianism, Abbott is certainly not alone in identifying dis- or unavowed metaphysics at the core of the social sciences, whether in sociology generally or these specific sub-disciplines. Nor, indeed, is he the first to question these assumptions. As will be demonstrated in this chapter, a number of researchers have discovered either overt or covert variants of them in their own areas of specialisation, either as unstated conditions of possibility of particular modes of enquiry, such as the GLM approach targeted by Abbott, or as more explicit commitments. These assumptions are, in many cases, species of the atemporality conjecture (or stand in close logical relations to it), a conjecture whose negation is in some cases a prerequisite for the validity of these theories. This research forces these assumptions to emerge from obscurity at the base of the relevant fields, highlights the need to question them, and opens up an alternative conception of the temporality of determinacy in the context of non-physical science. With this in mind, this chapter divides into several sections, each of which is dedicated to an extended analysis of one such field within the social sciences. Its goal is to demonstrate that such research is comprehensible only if one accepts the temporality of determinacy in a certain specific sense, and only if, in so doing, one accepts that time has an inalienable role in the corresponding laws.

Esposito: Endogeneity and Expectations

The economist Joan Robinson's collection of essays *What are the Questions?* provides a useful introductory perspective on the role of time in economic models. The logical time of equilibrium focused models corresponds to a "stationary state," in which economic relationships are preserved forever. In such a state, there is no evolution, but rather timelessness, and no distinction between different periods.[14] In contrast to this logical time of models, modelled reality is punctuated by disequilibrium, instability and unpredictable behaviour. So what is it that distinguishes these two times? For Robinson, the existence of decisions is the sufficient condition for time to take on a serious role in economic models. Agents make decisions in light of beliefs about their future consequences, and other agents' future decisions, that is, summarily, their expectations. Decisions made in light of inconsistent and uncertain expectations, later refined and revised, are the entry point of time.[15]

Although undeveloped and under-argued in her short but insightful essay "Time in Economic Theory," this connection between the appearance of agents' expectations in economic models and the temporal properties of the latter anticipates several modern theories. For the sociologist Elena Esposito, this connection is paramount. Given the fact that the systems economics concerns (more particularly, financial markets) are populated by agents who form expectations about an uncertain future, it is inevitable that it incorporates certain assumptions about time and, indeed, that these impregnate the discipline at such a foundational level as to prompt her to ask whether many of the *fundamental* contemporary problems of economics relate to its treatment of time.[16] For Esposito, some of the adverse consequences of the financialisation of the economy, such as endogenous crises, extreme volatility or the preponderance of speculation over investment, can be attributed to the way in which the future is viewed and managed and the way in which risk is referred to in the present,[17] and thus to beliefs which are implicitly or explicitly metaphysical.

The fact that Robinson and Esposito both take expectations to be driving forces in macroeconomic models is not in itself a heterodox theoretical anomaly. Keynes, in his *General Theory of Employment, Interest and Money*, for instance, accords expectations a prominent role in the context of his equation of aggregate investment with aggregate savings. He holds the disposition of individuals to spend or consume to be indissociable from others' incomes, ultimately linking expectations about expenditure and consumption to overall output: Levels of output and employment depend not only on production capacity or income, but on participants' expectations about consumption levels likely to exist in the future.[18] Keynes ranks expectations as being important enough to feature among the most vital conundrums which both impeded and inspired his writing.[19] But while expectations are clearly an important category in themselves in Keynes' theory, for Esposito and Robinson—as well as recent theorists such as Suhail Malik[20] and practitioners such as George Soros[21]—it is the temporal structure displayed by systems of expectations that is the most innovative feature for the purposes of this chapter. It is not expectations alone, but the relation between expectations and time, the claim that expectations imply extraordinary temporal dynamics, that is the more distinctive aspect of these two accounts, and which unlocks the most provocative theoretical conclusions of their work: time is in a sense the very *meaning* of the economy.[22]

In Esposito's case, these conclusions have extraordinary and perhaps grandiose implications. They include the production of indeterminacy in financial systems,[23] the recursive, specular or self-referential aspects of agents' expectations, and the rejection of asymmetry in the context of causal influences. Money first and foremost is used in order to manage the future's uncertainty and to produce contingency.[24] The weaknesses of economics arise from its failure to embed circularity into its analyses, and to grasp the inter-mixture of the present and the future.[25] Constraints and conditions which influence the development of an economic system are multidirectional rather than unidirectional.[26] Other examples of claims such as these can be found throughout Esposito's writings.

Regardless of whether one does or does not endorse these claims, as a theorist as native to the discipline as she is painstaking in teasing out its implications for the philosophy of time, it is natural to consider Esposito's research as important source material for any analysis of the temporality of determinacy in social science. This source material consists summarily of an extended argument against the atemporality conjecture. This argument rests on two contentious propositions. The first is Esposito's Endogeneity Thesis, which holds that time is a parameter specific to systems or sub-systems, in the sense that it is contextual and relative rather than absolute. The second is Esposito's Revisability Thesis, which concerns the formation and reformation of expectations (typically belonging to participants in financial markets). It is in this sense that she bears the heraldry of Keynes, Robinson and others in locating in the role of expectations the introduction of time to economic analysis. The conjunction of these theses, the theses of endogeneity and revisability, allow an argument to be constructed for the thesis that determinacy, considered in terms of the relations between the events affecting the system, is itself an ordered process. Accordingly, this argument is also an argument against sub-doctrines belonging to the atemporality conjecture.

Esposito's argument appears in recent works such as *The Future of Futures: The Time of Money in Financing and Society*, "The Time of Money in Finance and U.S. Society" and "Predicted Uncertainty: Volatility Calculus and the Indeterminacy of the Future." It can be reconstituted in the following way:

(1) Relations of time are internal to systems
(2) Expectations about prices comprise financial systems
(3) The future is bounded by the present future and future present
(4) The present future consists of expectations about prices and price movements
(5) The future is the revisability of the future present and the present future[27]

It is worth interrogating these premises one-by-one, before turning to the respects in which they contribute to the argument of this chapter.

The first is Esposito's Endogeneity Thesis (ET). Time is an endogenous parameter, belonging to the system it describes. How ought this claim to be understood? An indication as to her approach is provided in her chapter "Time Binding"[28] in *The Future of Futures*, where she refers to this as a principle of temporal relativity.[29] Esposito's presentation is minimalistic, given the imposing centuries of exchanges over absolute and relative notions of time in both the sciences and in metaphysics. The debates over absolute simultaneity between Newton, Clarke and Leibniz, the question as to the temporal *convention* which yields the simplest dynamical equations on the part of Poincaré and FitzGerald,[30] the post-Einsteinian absolutism as to space-time geometry, the modern formulations, such as that of Julian Barbour[31]—all these compose the relevant scientific backdrop to Esposito's discussion. She holds that temporal relativity (vicariously endogeneity) can be defined in opposition to a principle of absolute time, namely the view that time is an independent dimension, consisting of succeeding dates, determined in advance, with infinite extension.[32] Esposito this identifies several different aspects of absolutism: the autonomous and abstract nature of the time dimension, being independent of the events and processes which occur within it; the stability of this dimension, allowing any and all events to be referred back to it; and its infinity, understood as prolongation forward and backwards in this dimension. Although there are therefore several distinguishable principles lurking in this rather than only one, Esposito's presentation of the distinction between absolute and relative time seems reasonably faithful to the spirit of what can perhaps be regarded as its ultimate horizon, namely Newton's words in the *Principia*, according to which mathematical time or duration flows autonomously, independent of any external variable.[33]

By contrast, Esposito's characterisation of temporal relativity holds that time is a structure which belongs to systems, giving order to its operations.[34]

This is the most succinct expression of the ET: time is a structure or parameter of systems, that is, internal, thus endogenous, rather than external to them. The periodicity of the Earth's rotation, or indeed the oscillatory tilt of its axis, produces the alternations between day and night or the regularity of the seasons, and the corresponding cadence of anthropic activity such as waking and sleeping, or growing and harvesting crops. The decay of radioactive nuclei and the emission of such particles as neutrinos and electrons occur at rates specific to the isotope, expressed in statistical variables such as the mean lifetime and half-life. Celestial orbits, ontogenetic mechanisms, cellular differentiation and reproduction, right the way down to the processes we deem the most granular physically possible, such as ticks of atomic clocks, or the time of a photon's crossing a Planck length—all of these constitute their own temporal parameter by virtue of being stipulated as regular. This parameter is in this sense internal to each process. Moreover, there is a further element to Esposito's case: time provides an order for operations. This remark suggests a commitment on Esposito's part to an ordinal as opposed to cardinal, B-serial as opposed to A-serial framing of time. Each process (such as the ticking of an atomic clock) in no fundamental way measures the *extensity* of a given event; rather, the countable number of complete processes which occur (the number of ticks) in sequence provide a secondary measure of duration.

Construed more formally, Esposito's principle of temporal relativity has it that these different times, belonging to different systems and processes, need not be resolvable into one another. Adapting the terminology of Russell, consider the relationship between the times which serve to index some set of events and those which index another. Formalising this relationship requires a supplementary subscript with the $\{t_{1i}\}$ indexing the $\{e_{1i}\}$ and the $\{t_{2i}\}$ indexing the $\{e_{1i}\}$. Esposito denies that the times $\{t_{1i}\}$ which serve to index the events $\{e_{1i}\}$ can be reduced to $\{t_{2i}\}$ (whether by an identity function or mapping), denies the reverse, and, further, denies that whatever functional relation might be found obeying Russell's (modified) schema $E_t = f(e_{11}, t_{11}, e_{12}, t_{12}, \ldots e_{1n}, t_{1n}, t_1)$ holds both of $\{t_{1i}\}$, $\{e_{1i}\}$ and $\{t_{2i}\}$, $\{e_{1i}\}$. It is not the case that, given some time belonging to some system with its own dynamics, the same need be resolved into a superordinate time. This is the foundation for premise 1).

The second premise should now be familiar: financial systems are comprised of expectations about prices and their movements. Breaking with theorists' typical justifications and motivations for markets (especially securities markets) in terms of liquidity, efficiency, innovation, risk transfer and the enablement of hedging, Esposito's position limns the market as a theatrical environment in which unpredictably vacillating subjective or inter-subjective dispositions exert themselves, with severe repercussions, such as the permanentizing of financial crisis. In so doing, she repudiates the various methodologies which purport to approximate an objective determination of price, whether discounted cash flow models, appraisal-based approaches, or comparable transaction analysis, i.a. The various formulations of the efficient market hypothesis, for instance, as presented by Fama,[35] according to which the prices of securities reflect in full all information available about them, seem truly alien by comparison. Consider the strong, semi-strong and weak forms of the hypothesis which are separated in his work. Whereas the strong form contemplates whether individual actors or groups of market participants have privileged access to price-relevant information, the semi-strong form considers whether prices reflect all publically available such information, and the weak form restricts its focus to historical pricing information and return data in sequence.[36] Regardless of which is considered, there is quite some difference between on the one hand the vision of a market in which rational decisions taken in light of data—whether public information such as regulatory filings and financial statements or the historical returns on a security or index—determine pricing, and Esposito's vision, in which tenuous, second-order conjecture on other agents' perceptions and intentions are predominantly responsible.

Of course, the contention that expectations as opposed to rational decisions are relevant factors in appraising the dynamics of financial markets is not a unique one (nor one inconsistent with all forms of the EMH). The distinctive element of Esposito's focus on expectations is, rather, the reflexivity of the dynamics of these expectations. Esposito proposes the dependency, not only of prices on participants' expectations about future price movements but, more properly, these *expectations* on the expectations of other participants; forecasting future prices entails forecasting expectations, meaning expectations contemplate expectations. This constitutes the reflexivity of such expectations, as well as the specular and speculative nature of transacting on the markets they inhabit. In this respect, *The Future of Futures* cites the work of George Soros, who inscribes reflexivity into the essence of market economics: there is a reflexive qua two-way determinative relationship between the state of the market and the thought on the part of market participants about the market's state. These two dynamics interfere with one another, leading to feedback loops between understanding and reality. Soros explicitly opposes his view to the EMH, and to theories couched in terms of rational expectations.[37, 38, 39]

Next, Esposito's third and fourth premises open up another distinction, this time between two modes of time: the future present and the present future. For Esposito, the non-coincidence of the future price of a particular security or asset with the expected value of its future spot price is an expression of future uncertainty. The expected price of this security or asset so many weeks or months away need not equal the future price. The difference between the two captures the risk in the price of the security.[40] Esposito initially conveys this distinction in epistemic terms, attributing its provenance to the uncertainty of the future. Such an attribution at first risks trivializing premise 3), which would amount to a banality, insofar as the present future is the future one expects and the future present the one which obtains, stating little more than the truism that expectations sometimes prove inaccurate due to the informational constraints on market participants. Indeed, this is how Brassier reconstructs Esposito's argument, defining the difference in such a way as to suggest that the present future is anticipatory, whereas the future present is actual.[41] For his part, Brassier infers from this anticipatory presentation the presence of an unavowed commitment to idealism, which he takes to be a terminal weakness in Epsosito's account.[42]

Beneath this epistemic overlay there is, though, an ontological kernel. At this point in Esposito's argument, previous premises have established that the determinants of price levels, and changes in these levels, are expectations. This is reaffirmed by the definition given in premise 4). Insofar as expectations, as inter-subjective phenomena, elicit and orient transactions comprising the market, declarations about these expectations, and the effects they exert, entail idealism no more than talk of price movements fails to be objective talk. The contrast between the subjective world of expectations and the objective world of rule-governed prices (putatively functional on such data as traders' C-fibre activation, central bank monetary policy, the output of Monte Carlo simulations for price distributions, etc.), required in order for distinguishing between present future and future present in terms of the former to betray idealism or an exceptionable epistemic focus, does not apply to Esposito's framework (nor indeed the accusations of indeterminism which Brassier levels at Esposito on this basis). This much is clear from two other passages of hers featuring this distinction, in which the modes of time themselves take on a figurative power of action: the past and the future reconstruct the entirety of time from a particular perspective, Esposito claims, and there is a continuous process of production, revision and reconstruction of moments in time which leads to differing temporal integrations, different combinations of moments, and different relationships between forward-looking projections of the future and memories of the past.[43]

Taking into account all the passages which contain premises 3) and 4), then, what is the import of Esposito's distinction between present future and future present? To the extent that the future is "bounded" by present future and future present, it amounts to the interchange between these two modes of time: whereas the present future involves the formation of expectations (about future prices), the future present is the mode in which these expectations are *revised*. The future *redescribes* and *reconstructs* the expectations formed in the present future. The passage of time consists, then, not in the continuous prolongation of an autonomous dimension, but in an iterative, repetitive moment of revision.

Consider the following example. Take the present expected value of a security's spot price in three months. This is an expectation on the part of a market participant or group of market participants about the spot price three months in the future, that is, a present future. Compare this with the value of the security in three months' time, the future price, at a future present. An obvious question presents itself. Is the present expected value of the security's spot price in three months equal to its value, three months from now? Of course, it is common for them to differ materially—a difference which might naturally be ascribed to some set of financial and non-financial information (depending on one's stance with respect to the EMH), as does Brassier, for instance, when referring to such factors as traders' subjective mental states as explaining the market dynamics which dictate the time-evolution of price.[44] But, for Esposito, no combination of any of these factors can explain the differences between the expected value of the security and its future present value, nor any conjunction of them with any others. There is rather a minimum of (temporary) indeterminacy at work in this difference, which Esposito refers to as irreducible risk.[45]

At the outset, the attempt made by market participants to determine the expected value of the spot price in three months is, as before, a function of such data as traders' C-fibre activation, central bank monetary policy, or the output of Monte Carlo simulations for price distributions. This attempt to approximate the correct value produces a determinate quantity which influences the participants' decisions about the security. In a similar way, the future price of the security is a determinable quantity which one simply needs to read off the market value of the futures as they are being transacted (assuming the market is liquid enough; assuming it is not, some weighted measure or index of the same instead). But the *value* of the security at the time the expected value is calculated is *not* determined. It is not until three months have elapsed, not until the time of the future present, that the value of the spot price at this time, *contemplated three months prior*, is determined, even if a determinate approximation can be formed prior to this by the participant's intuition or calculation, and even if sundry forces belonging to many different systems might influence pricing dynamics.[46]

Esposito's contention is thus that this kind of indeterminacy and, accordingly, risk, is an intrinsic feature of the market. It is not a reflection of informational frictions which could be superseded in a fully transparent trading environment. The difference between the present future expected value and the future present spot price is a concrete manifestation of revisability: expectations about price movements are formed and reformed in time, until the determinate value progressively emerges.[47, 48] With each iteration, different combinations of factors are combined together and taken as grounds for the evaluation of the security's value, just as, prior to the endogenous financial crisis of 2008, historical returns may have buttressed the expected performance of now notorious asset classes which were rapidly devalued when widespread defaults emerged shortly afterwards.[49] This is the sense of Esposito's claim that present constraints are revised after being evaluated later on, in a distinct context.[50]

This additional dependency manifests itself in the interpretation of quantitative models for these securities. By way of illustration, the

Black-Scholes model (BSM) expresses the value of a call option in the following form:[51]

$$S_C(t) = S_0 N(d_1) - Ke^{-rT} N(d_2)$$

where $d_1 = \dfrac{\ln\left(S_0/K\right) + \left(r + \sigma^2/2\right)T}{\sigma\sqrt{T}}$, and $d_2 = \dfrac{\ln\left(S_0/K\right) + \left(r - \sigma^2/2\right)T}{\sigma\sqrt{T}}$

Here, the function N is the cumulative probability distribution function for a normal distribution, r the applicable discount rate, and K the strike price of the call option. Notionally the relation links the value of the option to the difference between the distributed future spot price and the strike price, adjusted for an underlying interest rate, and can be derived from parity relations, subject to a set of conditions. More important, though, is the presence of the volatility function σ, for which no *a priori* or observable distribution exists. Indeed, as many textbook presentations note and as Esposito analyses at length, it is common for market prices to be *assumed* to be objective measures of worth and implied volatility functions to be inferred using the BSM, rather than implied volatility being determined independently and used to assess the fidelity of market pricing relative to the model's output. Rather than this function determining the value of the call option in conjunction with the other parameters in the BSM, together with those others, it fluctuates from moment to moment, repeatedly shifting and readjusting on the unlikely promise of an ultimate valuation for a given time. It depends not so much on price fluctuations of the underlying but rather those of the derivative.[52] Esposito's identification of these aspects of the market's utilisation of the BSM accords with a recurring motif in her work, namely the circular interchange between the participants in markets and their reference assets (and, of course, between future and past, cause and effect),[53] here under the guise of implied volatility. The validity of the formula was in no way justified independently by its empirical use, nor were its predictive successes evidence that it was right; market volatility itself adjusted to the formula's estimates, according to Esposito. By influencing and shaping practitioners' expectations, the formula effectively confirmed itself.[54]

In *The Future of Futures*, the main asset classes which evidence the dynamic of revisability do indeed consist of derivatives, such as synthetic and securitised debt instruments or exotic combinations of options. Since the object of the (primarily speculative)[55] trade of such derivatives is frequently not so much the underlying asset nor even some linear function of its price, but rather the risk or volatility of its worth on the market, these trades confirm the difference between explicable, predictable, or determinable changes in price and the market's propensity for unaccountable turbulence. It is for this reason that Esposito takes derivatives to stand in a privileged relation to revisability, being used to manage and control the distinction between the future present and present future,[56] allowing participants to gain or reduce exposure to this risk.[57]

How can Esposito's stance, her notion of revisability, be formalised? Let $S_C(t)$ denote the spot price of a security C (such as a call option[58]) as a function of time, $\langle S_C(t_i)\rangle$ the expected value to some participant at the time t_i, and, accordingly, $\langle S_C(t_0 + 3)\rangle$ the expected value of the spot price of C at a time $t_0 + 3$ (in units of months). The implication is that t_0 is the time at which this expected value is, say, calculated as the output of a model. But for Esposito's account, this formalism is already insufficient. For her account debars time-independent expectation functions in the same sense in which it debars unrevisable such functions. The expected value $\langle S_C(t_0 + 3)\rangle$ does not only depend explicitly on the time of exchange and trading, indexed to the lived months of the trader, but also (albeit covertly) on the time at which this expectation is generated. More properly, we might introduce in addition to the $\{ t_i \}$ a set $\{ \tau_i \}$ whose members distinguish different revisions of the expectation function, denoted $\left(S_C\left(t_0 + 3\right)_{\tau_i} \right)$. Importantly, it need not hold that that $\{ \tau_i \}$ is a singleton, nor that for every pair $\tau_j, \tau_k \in \{\tau_i\}$, $S_C\left(t_0 + 3\right)_{\tau_j} = S_C\left(t_0 + 3\right)_{\tau_k}$. In this way, in order to capture revisability, the operator represented by these square brackets thus requires indexation no less than does t_i at which events e_i (here transactions or perhaps sets $e_i = \left\{ A_{e_i}, B_{e_i} \ldots, N_{e_i} \right\}$ of transactions) occur.[59]

What, then, is the relationship between these two sets $\{ \tau_i \}$ and $\{ t_i \}$, parameterising respectively the revisions of the expectation function and the activity of the trader? Moreover, what relation do they bear to the events $\{ e_i \}$ corresponding to the times $\{ t_i \}$? Are they functional upon one another, isomorphic, monotonic? Consider what mapping might obtain between $\{ \tau_i \}$ and $\{ t_i \}$. If this mapping maps the times over which the expectation function is determined to the times at which events $\{ e_i \}$ occur, it can be observed first of all that this map is *not* monotonic, in the following sense. Both $\{ \tau_i \}$ and $\{ t_i \}$ are ordered sets: the former consists of the times at which expectation functions concerning the prices of securities are formed and revised, and the latter of times at which events such as transactions occur. We can therefore write that $\tau = \tau(t)$, insofar as for the time of any transaction(s) there will be some expectation function, a function of t, indexed by a member of $\{ \tau_i \}$. It is not the case that for every $t_j, t_k \in \{t_i\}$ and $\tau_j, \tau_k \in \{\tau_i\}$, $t_j \geq t_k \rightarrow \tau_j \geq \tau_k$; there will be some pairs τ_j, τ_k such that $\tau_j < \tau_k$ despite that $t_j \geq t_k$. The initial assessment of the value of the future spot price takes place at t_0 which is before the ensuing three months, yet the expected value of the spot price $\langle S_C(t_0 + 3) \rangle$ is determined *after* the revisions to the expectation function are effected—at, say, the τ corresponding to $t_0 + 3 + \delta$. Thus $t_0 < t_0 + 3 + \delta$ yet $\tau(t_0) > \tau(t_0 + 3 + \delta)$ for $\delta \geq 0$. This is the sense in which whatever relation may obtain between the $\{ \tau_i \}$ and $\{ t_i \}$ is not monotonic or order-preserving.

Whilst the repercussions of these accounts for the temporality of determinacy are likely already discernible, how can these be articulated more clearly? Recall the definitions given by Russell of deterministic systems and, subsequently, the Truth-Value Constancy (TVC) interpretation of the determinacy of events subject to a law f:

> A system is deterministic if, and only if, for events $e_1, e_2, \ldots, e_n$ occurring at times $t_1, t_2, \ldots, t_n$ and for any state E_t of this system at any time t, there exists a function relating the $e_1, e_2, \ldots, e_n$, $E_t = f(e_1, t_1, e_2, t_2, \ldots e_n, t_n, t)$.[60] (Def. Determinism)
>
> For any law f, and any occurrence of any $t_l \in \{t_i\}$, the truth-value of f does not change with variation in t_l (Truth-Value Constancy Thesis)

How does the conjunction of these two sit in view of Esposito's theses? The TVC thesis can be seen to fail quite directly, in as much as changes in t involve changes in τ, and in as much as changes in τ change not only the truth-value of expectation functions but also the syntax and content of such functions. Specifically, in Esposito's model, where ° denotes the truth value of the preceding function:[61]

$$\left(S_C\left(t_0+3\right)_{\tau(t_0)}^{\circ} \neq S_C\left(t_0+3\right)_{\tau(t_0+3+\delta)}^{\circ}\right)$$

To the extent that this expectation function f is a candidate for a law describing the price movements over time of this security which ought to obtain, we have a case in which a variation $t_0 \rightarrow t_0 + 3 + \delta$ changes the truth-value of f.

These considerations establish the following conclusions. First of all, the *determination* of events, understood in line with the previous chapter's analysis as their subjection to some mathematical function (for instance, the subjection of the pricing of a call option to the BSM formula), is a process, in the sense that it admits of an order. Second of all, there is a distinction inherent in Esposito's work and the kinds of explanations it poses, between the order of events ($\{ t_i \}$), and the order in which events are determined ($\{ \tau_i \}$). These two features delineate the impact of her theory on the question as to the temporality of determinacy—an impact which is stark. It is not simply that different information is available to market participants at different times, and that the appropriate price of a security is subject to revisions from time to time. This would be to posit only asymmetric relations of accessibility between moments: as time marches on, more data becomes available, and the wisdom of hindsight enables past follies to be corrected. Esposito goes a step further: establishing of an order of events, and subjecting these events to the functions which describe them, are temporal and processual.

Esposito clearly herself perceives the latent generality of her account. In the transition from ancient to modern societies, she conjectures, the vision of a static temporal order, located then in the divine perspective on *aeternitas*, dropped away, and the uncertainties represented by time became intrinsic rather than the results of the finitude of human cognition.[62] The eternity, here understood as fixity, of the ultimate order of things constitutes the atemporality of the determinacy Esposito attributes to the ancient conception of the world. The open future she puts up as an alternative dispenses with such an ultimate order, wresting time from its perspectival aspect, and reinvigorating it as a recursive form.[63] The fabric of time itself is disrupted by the inter-dependency of expectations and the precipitous revisions subsequently brought to bear on the laws of pricing. As will be demonstrated in this chapter, conceiving a determination process, an order in which events are determined, to be no less a feature of the evolution of the world than is the order in which events occur, is a commonality shared with the research of other scholars in the social sciences, both in the adjacent terrain of statistical finance, and research in the political sciences, such as that of Brian Massumi, as well as in recent technological innovations.

Time Series Analysis

Unique as is Esposito's intervention in the foundations of finance—an intervention which finds in these foundations significant volumes of disguised metaphysical content—it is not entirely unprecedented. Recent years have seen a detectable increase in the volume of research undertaken in this area. Its principal themes have been those such as the co-dependency of theory and practice and the contingency of the market, conceived as sequences of prices for securities—themes to which Esposito alludes also. However, it has also generated discourse which impinges much more closely on the nexus of temporality and determinacy. Timothy Johnson's paper "The necessity of multi-disciplinary scholarship for finance" relates developments in the field to the immutability of physical laws, reminiscent of the way in which the permanence of laws features as an essential kernel in Lindsay and Margenau's work decades prior: Financial mathematics cannot simply rely on immutable laws or lack of evolution in natural environments.[64] Jon Roffe dedicates a chapter of his 2015 *Abstract Market Theory* to the temporality of markets, or, in his terms, the temporality implied by the orthodox understanding of market dynamics,[65] drawing in the course of his discussion on Hume, Meillassoux, Deleuze, Foucault, Lacan, and Freud. In parts, Roffe goes as far as to deny the possibility of prediction of future states.[66] Suhail Malik, in his contribution to the 2014 *Collapse VIII*, follows Roffe in his focus on aspects of the market's contingency: time enables systems to represent their own contingency.[67] For Malik, this in turn entails a non-random[68] indeterminacy of sorts, securing the link between the temporal characteristics of the market's dynamics on the one hand and (in)determinacy on the other. Elie Ayache and Donald MacKenzie draw analogous connections.[69, 70]

It should come as no surprise that this increasingly rich tapestry of research into the nexus of temporality and determinacy has arisen in quantitative finance, economics, econometrics and related disciplines. It is observable also from the statistical methods and techniques they employ, from, for example, their prominent use of time series analysis. This discipline will be the focus of this section, along with its implications for the temporality of functional relations. Although far from confined to this list of subjects and equally present in others such as engineering or epidemiology, techniques developed to handle time series data (data with a temporal or sequential structure) have a particularly central role in these disciplines. There is no paucity of examples. James Hamilton's seminal work in the field studies various applications of time series analysis, from its use in forecasting macroeconomic variables such as money demand and elasticity to stockholder satisfaction or utility.[71] Academic and professional programmes of study emphasise the importance of regressions of time series data for the modelling of the financial markets' dynamics, considering their relevance to a range of securities classes from equities to more exotic derivatives. Campbell, Lo and MacKinlay in their work *The Econometrics of Financial Markets* frame the performance of an asset class or market as a series of returns indexed by time, returns which are in turn functions of prices so indexed, so that $R_t = \frac{P_t}{P_{t-1}} - 1$.[72] William Brock's extensive writings bring time series approaches to bear on everything from appraisals of the outperformance of technical analysis in trading, to the Cobb-Douglas production function, to the relationship between these financial models and statistical mechanics, chaos and nonlinear dynamics.[73]

Considered in themselves, the time series data which are studied in finance do not differ fundamentally from the data presented by any other system. So, *a priori*, any series of observations over time might be amenable to the Russellian framework of sets of events $\{e_i\}$ effectively indexed by times $\{t_i\}$. It is instructive to consider how the formal characteristics of time series and their models compare with the descriptions of functional relations given by atemporality theses, providing a useful data point for testing the working hypothesis of this chapter. In this new context, the explicit reference to a set of times is dropped in favour of a set of observations or values $\{y_i\}$. Thus, Hamilton, in his *Time Series Analysis*, defines the series as a group of observations with the date of each serving as an index, $\{y_t\}_{t=-\infty}^{\infty}$.[74] As a basic case, he considers first-order differential equations (instances of difference equations), to take (in their basic manifestation) the form $y_t = \phi y_{t-1} + w_t$,[75] in which each successive state is linearly related by some constant ϕ up to a function w_t. Hamilton considers subsequently many more sophisticated dependencies between observables and their lagged values, not least those obeying the more general form $y_t = \phi_1 y_{t-1} + \phi_2 y_{t-2} + \ldots + \phi_p y_{t-p} + w_t$. Going further, an array of techniques have been developed for the modelling and forecasting with time series analysis, as well as an array of classifications.

In order to grasp the meaning of these techniques, some rudimentary distinctions and properties ought to be introduced. Commentators distinguish between univariate and multivariate time series. The former consist of observations of a single variable, the latter of those of several. Some commentators bring up an additional distinction between regular time series, which are observed or collected at pre-determined intervals, and irregular time series which are not (a distinction which poses clear problems for anyone committed to a principle of endogeneity). On the one hand, the time points which parameterise the observations can be discrete or, on the other hand, continuous. But perhaps the most salient distinction presented in foundational accounts of this mode of analysis is the distinction between stationary and non-stationary systems. Whereas in the case of the former, the core statistical parameters in a given relationship, such as the means or variances of the values $\{y_i\}$ are constant or "stationary" over time, the latter examples violate this condition. Following Hamilton, if the mean μ and the autocovariances y_{jt} are independent of the date t, then Y_t qualifies as *covariance-stationary* or *weakly stationary*: thus $E(Y_t) = \mu$ for every t, $E(Y_t - \mu)(Y_{t-j} - \mu) = \gamma_j$ for every t and any j.[76] Alternatively, in the terms of Chatfield and Xing, stationary time series are those which exhibit no systematic change in mean, nor variance, and without periodic changes, meaning the system in effect possesses similar properties over time.[77] More formally, a time series consisting of values of a random variable $X(t)$ or X_t is said to be strictly stationary if $X(t_1), \ldots, X(t_k)$ is the same as $(t_1 + \tau), \ldots, X(t_k + \tau)$ for all $t_1, \ldots, t_k, \tau$,[78] *mutatis mutandis* for X_t. The latter is a more restrictive condition than the heuristic explanation.

So, introductions aside, what fecundity does this discipline provide for the questions posed as to the temporality of determinacy, and how and why might it impact the atemporality conjecture? The question to be posed of these models of time series data, which enables an enquiry to be made into aspects of determinacy, concerns the existence and nature of the functional relationships they comprise. More specifically, it concerns whether and how a function f fulfilling Russell's schema appears in applications of time series analysis to data—the schema, that is, which allows the determination of all members of $\{e_i\}$ or, in this context, all observations $\{y_i\}$, given (i) f and (ii) a suitable specification of initial conditions. The various functional forms which can be applied to model a time series, and the innumerable values of the ϕ_i which are substituted in order to improve the fit, are clearly attempts to find such a function.

It is noteworthy, however, that the frequent failure to find a sufficiently close model for the $\{y_i\}$ has led the notion of exact fit to be replaced in applications of time series analysis by notions of statistical significance. It is accepted by practitioners that, even with the endless permutation and combination of coefficients and sundry elaborate variations in functional form, only an approximation can be attained. This is to say that it is not the purpose of the theory to specify an exact $E_t = f(e_1, t_1, e_2, t_2, \ldots e_n, t_n, t)$. Rather the data $\{y_i\}$ may not match the constituents of the y_t, hence the need to include the residual terms w_t to resolve the mismatch, and the incompossible perspectives on the status of them. An abundance of tests, such as t-statistic tests, are employed to discriminate between degrees of statistical significance, based on the data's assumed conformity to *a priori* distributions. The adequacy of a model is quantified by a series of tests: its negation, that is, the hypothesis of negligible statistical significance or "null hypothesis," is compared against a sample of time series data using a test statistic. A refutation of the null hypothesis occurs strictly relative to a predetermined threshold of significance, which is to say that the validity of the model tested holds only within a defined confidence interval—the initial implication being that finding anywhere near exact functions is altogether rare.

The differences between observations and constituents of the model (the $\{y_i\}$ and the $\phi_1 y_{t-1} + \phi_2 y_{t-2} + \dots + \phi_p y_{t-p}$) are, in part, captured by error terms which reconcile the two. These errors are the subject of many different models and are taken as manifestations sometimes of a stochastic process, sometimes a deterministic one. Hamilton, for instance, distinguishes between two interpretations of this part of his first order equation, taking it as a random variable or as a sequence of numbers which are deterministic.[79] He proceeds to give an extended treatment of the stochastic interpretation, the first of these two. This treatment of the residuals means that the process itself, described by y_t, is a stochastic function, regardless of the nature of the elements $y_{t-1}, y_{t-2}, \dots, y_{t-p}$ which comprise the right hand side of this linear equation.

What is the significance of the stochasticity of time series descriptions—the stochasticity evident from the absence of exact descriptions from its models, and the presence of stochastic error terms? Does this enable conclusions to be drawn right away as to the temporal status of laws? Chatfield and Xing, in their *Analysis of Time Series*, reflect on this question, providing an exordium before their technical analysis of stochasticity which encompasses random walks, random walks with drifts, moving average processes and purely random processes. Queue lengths, road traffic accident frequencies, air temperatures at a particular place over the course of several days—all these are styled as chancy phenomena, constituting stochastic systems, governed by probabilistic laws.[80] Doubtless, these remarks venture beyond the temporality of determinacy and into the precipitous terrain of determinism and indeterminism proper. Stochastic processes are here defined in opposition to deterministic processes, and equated with randomness in a way which violates the foundations of determinism. However, there is no contention that determinacy requires determinism, any more than there is the contention that whatever variables of interest are never determinate at any time, even after having occurred. After all, determinacy holds of every random event upon its occurrence. So there is no reason to think that this soft commitment to indeterminism precludes this extract from offering insight into the question of determinacy. Reinforcing this point, the authors take probabilistic laws to govern the evolution of the system in question over time. In as much as the question as to the nature of these laws (*qua* mathematical functions) is embedded in the various forms of the atemporality conjecture, as well as being an important component of the very meaning of determinacy, whatever the methodology brought in time series analysis should be taken into account. In this sense, there ought to be nothing disconcerting about the fact that stochastic processes, embedded as these accounts show them to be in the very groundwork of the subject matter of time series analysis, have implications for the temporality of determinacy.

What, then, might these implications be? Processes such as those described by the likes of Chatfield and Xing re-interpret the functional relations describing physical or non-physical processes, substituting in place of relations which fix or determine events arbitrarily displaced in either direction in time functions which yield probabilities for said events.[81] Accordingly, the most relevant question as to the implications of this framing of stochastic processes is less its relationship with definitions of determinism, and more its impact on determinacy: regarding them as a group of random variables defined at a group of times, and taking "defined" as broadly coterminous with "determinate" provides a clear sense in which determinacy is temporal or processual, insofar as it suggests that the time points at which events occur are the time points at which events are determined, each being random processes. Although contrary to Esposito's posit of a thoroughgoing *distinction* between the events ordered and the ordering process, or, more properly, between the order in which events occur and the order in which they are determined, it aligns with her position that determination is itself a process—provided that determination is taken to refer to the outcome of random trials, rather than to the subjection of events to deterministic functions. For now, then, the clear commitment to a basic kind of temporal determinacy can be noted, due to the fact that events *become* defined or determinate in the same way that they *become*.[82]

As an example, consider the class of strictly or purely random processes described by Chatfield and Xing, being discrete-time processes consisting of sequences of independent random variables (tantamount to observables). These processes in fact fulfil the definition of strict stationarity, having a mean and variance constant over time. Particularly noteworthy is the case of the autocovariance function—the covariance function applied to random variables—defined in the following way:

$$\gamma(t_{1,}t_2) = E\{[X(t_1) - \mu(t_1)][X(t_2) - \mu(t_2)]\}$$

This is in effect a generalization of the variance function, a standard measure of the dispersion of a distribution of random variables, to situations in which the correlation of two variables in a distribution is considered. (Alternatively, the variance could be considered a special case of the autocovariance function for the case in which $t_1 = t_2$.)[83] Consider then the autocovariance function $\gamma(\tau) = Cov[X(t), X(t + \tau)]$, that is, the autocorrelation function depending on the difference $t_2 - t_1$, namely, the lag. This lag displays interesting properties for the case of purely random processes; in particular,[84]

$$\gamma(k) = \operatorname{cov}(Z_t, Z_{t+k}) = \begin{cases} \sigma_k^2, \; k = 0 \\ 0, \; k = \pm 1, \pm 2, \ldots \end{cases}$$

and also $\rho(k) = \begin{cases} 1, \; k = 0 \\ 0, \; k = \pm 1, \; \pm 2, \ldots \end{cases}$ where $\rho(k) = \dfrac{\gamma(k)}{\gamma(0)}$.

These formulae express formally the following proposition: *Arbitrary pairs of distinct variables have no correlation with one another* and are in this sense independent. As will be examined more thoroughly in the next chapter, which examines the temporality of determinacy in the philosophy of physical sciences, this criterion for the independence of random variables parallels the notion of probabilistic irrelevance which arises in the context of ergodic, stochastic and Bernoulli physical systems, and for which $P(e_l | e_k) = P(e_l)$ for a pair of events $\{e_k, e_l\}$.[85] In either case, the connection between (i) the conjunction of various initial conditions (*qua* determinate events, or the outcome of a trial of a random variable) with whatever rules or laws govern their evolution—be they probabilistic or not—and (ii) other occurrences, is severed. If this is accepted, the determinacy of these events or random variables is accordingly as temporal as are the events themselves.

The stochastic properties of functions such as y_t which follow from the nature of their residual terms w_t is by no means the only appearance of temporal determinacy in the analysis of time series. Another is autocorrelation, or serial correlation. These words denote the tendency for the subject dependent variable to be correlated with a lagged version of itself, or with multiple lagged versions of itself. Such models can be deemed autoregressive (AR) models, and are ascribed an order corresponding to the number of lagged variables included in them as independent variables. Statistical tests such as the Durbin-Watson test or t-test can be employed to measure levels of such serial correlation. The existence of strong serial correlation or, indeed, significant correlation among the residual terms of the AR model (seeing as the value of these error terms w_t equally vary with time) can limit the predictive or explanatory power of the model. A further property, heteroskedasticity, defined as the inconstancy of the variance of the error terms or residuals themselves over different data in the time series, is closely related, and is often addressed by deploying autoregressive conditional heteroskedasticity (ARCH) models. Just as non-stationary behaviour involves the variability of models' fundamental properties and stochastic behaviour temporalizes the determination process (by collapsing it onto occurrences or the outcomes of trials), the intrinsic presence of time variables in their functional forms suggests a meaningful time-dependency exists in their very syntax.

Most importantly, how can these observations drawn from time series analysis—concerning such features as stochasticity, stationarity, serial correlation—be used to confirm, refute or otherwise elucidate the atemporality conjectures? To facilitate this, it is convenient to recapitulate the sub-doctrines set out in previous chapters:

> Under the mapping $t \rightarrow t - t_0$ solutions of dynamical equations map to solutions of dynamical equations. (Translation Invariance Thesis)

For any law f, and any occurrence of any $t_l \in \{t_i\}$, the truth-value of f does not change with variation in t_l (Truth-Value Constancy Thesis)

For any law f, f is permanent (Permanence Thesis)

For no law f does the time-variable t occur in f where $dg'' + bg' + cg + d = f$ and $g' = \frac{\partial g}{\partial t}$, $g'' = \frac{\partial^2 g}{\partial t^2}$. (Time-Variable Thesis)

First of all, consider these theses in light of the property analysed above in the case of stochastic models, namely the independence (related to probabilistic irrelevance) of pairs of variables, trial outcomes or observations separated by a temporal lag. These theses concern functional relations or laws f, whose time series analogue is simply the model in terms of which observations are related to one another. Now, insofar as each of the atemporality theses sets out a candidate necessary and sufficient condition for a valid or legitimate rule or law taking the form of a functional relationship, examples of functional relations which fail to befit such conditions amount to refutations. Yet there are serious challenges squaring any of these conditions with a stochastic system which consists of a set of trials of random variables $\{Z_t\}$, which can be readily illuminated by two extremely straightforward examples of functional relations governing the outcomes of these variables: Bernoulli and Poisson distributions. Grimmett and Stirzaker in their *Probability and Random Processes* provide accessible presentations of these distributions.

Binomial Distribution. A coin is tossed n times. For each toss, there is a probability of attaining a head $p = 1 - q$. Define $\Omega = \{H, T\}^n$, and X as the number of heads, taking values 0, 1, 2, …, n, being a discrete random variable. The associated probability function $f(x) = P(X = x)$ is such that $f(x) = 0$ if it is false that $x \in \{0, 1, 2, \ldots, n\}$. If $\binom{n}{k}$ points in Ω yield k heads in total, the probability of attaining each of these points is defined by $p^k q^{n-k}$, meaning $f(k) = \binom{n}{k} p^k q^{n-k}$ if $0 \leq k \leq n$. This is the meaning of the claim that X exhibits a *binomial distribution* with parameters n, p.

Poisson Distribution. Assume a random variable X takes values 0, 1, 2, … with a probability function. $f(x) = \frac{a^k}{k!} e^{-a}$, $k = 0, 1, 2, \ldots$ for $a > 0$. X follows the *Poisson distribution* with parameter a.[86]

Some of the as yet unmentioned terms appearing in these definitions, conceivably liable to cause confusion, can be resolved into more familiar vernacular. Probability mass functions are functions which stipulate the probabilities associated with individual values (outcomes) of the random variable being considered or ranges of such values.[87] The symbol Ω represents the set of different outcomes, while $P(X = x)$ is the probability associated with the member x of the set X ascribed to the random variable. In such cases, these functions are the "rule" or "law" that may be thought to correspond to the f which is the subject of the atemporality conjectures: rather than determinative dynamical equations, the functions have a statistical meaning, employed in forming predictions as to the outcomes of trials, or, more accurately, the distribution of outcomes expected for a sufficiently great ensemble of trials. On the face of it, aspects of various of the forms of the atemporality conjecture are indeed concordant with these simple models: the time-variable appears in neither, boosts in time would appear to have no effect, and the generality of the Bernoulli and Poisson distribution functions, extending as they do to arbitrary n, could be taken to underwrite their permanence as putative laws. However, the functions f appearing in the guise of these distribution functions differ markedly from those relations embedded in the conjectures. Recall the description of the functional relations introduced previously in the latter context:

> A system is deterministic if, and only if, for events $e_1, e_2, \ldots, e_n$ occurring at times $t_1, t_2, \ldots, t_n$ and for any state E_t of this system at any time t, there exists a function relating the $e_1, e_2, \ldots, e_n$, $E_t = f(e_1, t_1, e_2, t_2, \ldots e_n, t_n, t)$.[88] (Def. Determinism)

Although Russell's deflation of the notion of determinism into functions might be unexpectedly satisfied by functions which express only probabilistic relations, it can reasonably be taken as an implicit stipulation that the occurrence of $\{e_i\}$ in the schema he provides, namely $E_t = f(e_1, t_1, e_2, t_2, \ldots e_n, t_n, t)$, are required to be determinate if subjected to f. This stipulation would disqualify probability distributions from conferring determinacy on the events distributed. On the basis of this reasonable assumption, the difference between the deterministic laws f considered in the atemporality conjectures on the one hand and probabilistic functions brought to bear on stochastic processes on the other becomes dramatic, and the atemporality these would be connected to time series analysis in a far less natural way.

Second of all, putting aside the question as to the implications of stochasticity, turning instead to the non-stationarity of many of the distributions in question, a serious clash can be observed between this property and the notion of permanence. By way of a basic sketch, following Hamilton's definition of stationarity given above,[89] for non-stationary systems it fails to hold that $E(Y_t) = \mu$ for all t with μ taken to be a constant. Insofar as the permanence of the "law" f or, in this context, the time-series description Y_t, requires the permanence of the expectation function of f, and insofar as this permanence requires in turn its invariance with t, the existence of non-stationarity suggests that these models repudiate this instantiation of the atemporality conjecture. Equally, phenomena such as the heteroskedasticity of error terms or auto- or serial correlation provide a supplementary illustration of core properties varying in ways which reconcile poorly with permanence; the variability of the error terms involved in a given model itself varying over time can hardly be held to befit models or descriptions with sempiternal essential features.

Putting permanence aside, what of the other sub-doctrines of the atemporality conjecture—such as the TVT, or TVC? An almost trivial observation is that the actual series being modelled here very explicitly feature occurrences of time variables; in indexing the observations, they constitute the series, which is to say, the differences between elements, and are therefore fundamental to the methodology. A time series free of time variables would of course be straightforwardly self-abrogating, hence the potential for the discipline to impinge upon the viability of the Time-Variable Thesis (TVT). In its cruder rendition this thesis debars wholesale the appearance of time-variables in the relevant equations, a preclusion which is trivially incompatible with almost any practicable model, and certainly the models proffered by Hamilton, subject only to highly artificial exceptions. But what of the more sophisticated, charitable reading of this thesis contributed by Lindsay, Margenau, Ushenko *et al.*? The focus on second-order differential equations as a proxy for dynamical laws is, as observed above, rather foreign to the techniques and practices of time series analysis in which the functions typically have a statistical or aleatory meaning. Nonetheless it is possible to facilitate a comparison between the two, or at least between the spirit which motivates one and that of the other. The kernel of these authors' insight is that the appearance of time in the dynamical equations which encapsulate the behaviour of the system reflects the changes to the system brought about by the forces which impel it; these forces, captured by functions f, do not themselves contain t. This motivates the separation of the function g whose derivatives populate the left-hand-side of the formula which appears in the TVT from the function f which describes these dynamics. f's dependencies should be spatial separations, charges, gravitational constants—not temporal parameters. This, for Lindsay and Margenau, simply reflects the axiomatic uniformity of nature. Thus, for instance, they discuss how the factors which appear in Coulomb's law of attraction should take the same value forever—so for all parameters in the differential equations which they believe to encode the laws of nature.[90] But this line of thought is surely doomed to failure once the considerations laid out in this section are fully absorbed. The paradigmatic functional forms describing time series, such as difference equations replete with lagged variables, involve sequential temporal dependencies. Yet such functions are precisely the analogues in the time series arena of those functions f held to be time-independent by proponents of the atemporality conjecture.

In respect of the Truth-Value Constancy Thesis (TVC), trivial counterexamples can be generated. Take simply the general form of the autoregressive series $X_t = \alpha_1 X_{t-1} + \ldots + \alpha_p X_{t-p} + Z_t$ for $p = 1$, so that $X_t = \alpha_1 X_{t-1} + Z_t$, a simple first-order autoregressive model which might describe a phase of rapid economic expansion, or the infection count of a novel communicable disease. Consider the triple of time series elements $\{t_1, t_2, t_3\}$ and the valid formula $X_3 = \alpha_1 X_2 + Z_3$ generated by substituting $t = t_3$ into the series. Under the mapping $t_3 \rightarrow t_1$ the truth value of the model need not be constant; for $t_1 \neq t_3$ and for nontrivial values of α and Z one can easily convert legitimate formulae into illegitimate ones.

A comparable lot befalls translation invariance. Although (somewhat like the TVT) "solutions of dynamical equations," have no obvious counterparts in the context of time series, observations or random variables related by whatever model qualify as a reasonable approximations of them. On this assumption, one can make a trial with this formulation of the atemporality conjecture by considering whether maps between time series data—substitutions of data for data—preserve or violate relations between the model in question. It can readily be observed that autoregressive series such as $X_t = \alpha_1 X_{t-1} + \ldots + \alpha_p X_{t-p} + Z_t$ fail to preserve the requisite relationships under the mapping $t \rightarrow t_0$ for at least some suitably chosen t_0, in an analogous fashion to that in which the TVC can be shown to fail for basic AR(1) models.

These observations summarise the implications of time series analysis for the temporality of determinacy. Understanding these requires first of all considering the possibility of finding functions which relate events in time to other events in time in the context of time series analysis, despite the deep embedment of statistical meaning in the subject models, which is especially discernible in the presence of residual or error terms. Second of all, though, it requires an analysis of the status of these functions in relation to species of the atemporality conjecture, concerning permanence, truth-value constancy, translation invariance and the appearance of the time-variable. From both of these perspectives, it is difficult to avoid the conclusion that the atemporality conjecture reconciles very poorly with the direction and character of research in this domain—which consists intrinsically in the development of statistical techniques to handle data, defined in each and every case by their temporal location, embedding time-dependency in the essence of its descriptions.

Massumi: Temporality of Pre-emptive Logic

Whilst much of the West's interventionist military doctrine of the early 2000s has by now been anathematized in academic literature, few accusations have been cast at the would-be logic, or illogic, operative in its discourse. For a small but growing number of recent commentators, though, most notably Brian Massumi, this line of enquiry is essential for the production of a complete empirical account of the events comprising the "war on terror," and also reveals the extent of the conceptual gymnastics performed in the service of the economic, political and ideological interests motivating the intervention—interests to which ontology is subordinated and weaponized as a malleable discourse, as opposed to being their *a priori* intellectual foundation. Thus, just as Esposito finds in the theory and practice of finance an unavowed set of commitments to dogmas in the philosophy of time, so does Massumi in the distinct arena of the militaristic (il)logic of threat and pre-emption, drawing connections to the ideas of Kierkegaard, Deleuze, Nietzsche, Whitehead and C. S. Peirce, and declares that time is itself a political issue.[91]

The notion of *pre-emption* is the key to Massumi's account. By way of source material he draws on everything from political speeches (such as those of George Bush preceding the war in Iraq) to academic apologists for intervention. This notion is by no means limited to a purely military meaning. Massumi argues that it in fact distinguishes a particular *logic of power*, definitive of an epoch of international relations, with both ontological and epistemological implications.[92] Thus, the continual references in his writings such as *Ontopower* to the *logic* of pre-emption should not deceive the reader into thinking that the proximate philosophical implications of this emerging discourse are confined to such areas as propositional calculus, rules of inference, or questions of predication and quantification. Rather, this logic is inter-woven with wider metaphysical questions, particularly questions of temporality: the metaphysical status of different modes of time, the coherence or conceivability of future causes and the relationship between epistemic uncertainty and ontological indeterminacy. This section will distinguish two main aspects of Massumi's conceptualization of pre-emption. First of all, it presents the temporal logic of the future threat, the object of pre-emption, which he explains in the form of a conditional: Actor 1 performed action A because Actor 2 could have performed something which Actor 2 did not, in fact, do.[93] Second of all, it exposits Massumi's conception of future causality, that is, his claim that events can stand in causal, *qua* determinative, relations with events which precede them. This claim entails not only a principle also recognizable in Esposito's research, namely the distinction between the order of events and the order in which they are determined, but a further, more radical affirmation of the outright reversal of these two orders: an event occurring later in time nevertheless determines its predecessor—hence Massumi's liberal employment of motifs of circularity, self-reference, recursion and feedback.[94]

The following statements summarize Massumi's analysis of the logic of pre-emption. There are a number of distinctive elements to be analysed: (i) the modality of the formulae describing the threat, (ii) the denial of noncontradiction between competing states of the threat, (iii) the analyticity of the descriptive statements used to underwrite the associated moral imperative.

> Pre-emption escapes laws of classical logic such as non-contradiction, as well as linear, one-way causal relations between past and future. Instead, it describes affect, which occurs in a non-linear time where a recursive relationship holds between future and past.[95]
>
> Threats are signs brought about by future causes.[96] The threat's potential can be represented only by a conditional logic, which describes a potential rather than a reality. Such statements, concerning what was not, but could have been, are indefeasible, and unverifiable.[97]
>
> Speculative acts of pre-emption *create* the adversaries they aim to counter-act. Pre-emption does not attempt to counter-balance a threat or prevent an attack; it pre-empts the future and in so doing, creates the threat whose existence it speculates about.[98]

In respect of (ii), Massumi is clear that a threat actor's activity eludes the principle of non-contradiction: the threat is constituted in neither a determinate state (of e.g. incipient attack, possession of certain weapons, etc.) nor the negation of this state. Statements about the intentions and machinations of an adversary, or about the suppositious modalities of permanently latent conflict, can be in some sense both true and false. In these passages, Massumi links this peculiarity to the altered notions of causality and futurity which prevail in the dynamics of pre-emption. As for (iii), the analyticity of the descriptive statements about the adversary is a clear commitment, with Massumi denying that conditional statements of the sort espoused by interventionists can be wrong. He claims interventions are right by definition, analytically,[99] in referring to contingent future events as being possible, or in making self-verifying declarations about one's perceptions of threats.

However, the most distinctive and most relevant aspect of Massumi's *Ontopower* for the question of time is thesis (i), which focuses on the modality of the formulae describing the threat. Massumi distinguishes several possible types of modal statements which might best account for the pre-emptions witnessed in recent theatres of conflict. According to one, threats are actual, empirical, extant, and determinate states of affairs. To identify such a threat is nothing other than to make an empirical statement amenable to the logical form Pa (or a conjunction thereof) true at a given time. Such threats are acts already in progress. This interpretation, however, suffers from a defect; it neglects the anticipatory character of pre-emption: to respond to an attack already occurring is, of course, to pre-empt nothing; it is rather to react and defend oneself afterwards. Alternatively, a second straightforward interpretation is that such a threat is rather a *possible* empirical, extant, and determinate state of affairs, a possibility perhaps tied to futurity. This would be formalised by the sentence $\Diamond Pa$ (or by a conjunction of similar atomic propositions). For Massumi, though, this formulation neglects the specular aspect of pre-emptive action. From his perspective, they have a second-order, iterated modality,[100] giving (controversial) meaning to the possibility or necessity *of a possibility or necessity*. The object of pre-emption is neither a present possibility nor a future actuality, but rather a future possibility, an eventuality that may *emerge as possible*, and thus an object even further unmoored from actuality: $\Diamond \Diamond Pa$.

What evidence can Massumi proffer that this peculiar modal status was indeed one belonging to the cited discourse, that is, advocates and apologists for these recent conflicts? First of all, he cites George Bush's admission that military action was justifiable to dispel non-existent threats: threats must be confronted before they even emerge as threats.[101] When questioned, Bush justified his actions *ex post facto*, retrospectively, based on vague assertions. In his assessment of the results of the war on terror, the mere capacity to have produced threatening weapons of mass destruction was transformed into a threat: the capacity could have been used to produce weapons, such weapons could have passed into the hands of those willing to use them.[102] Finally, once knowledge of the actual capabilities of the adversary at the time of pre-emption is attained—knowledge acquired long after this pre-emption—the falsity of empirical statements as to the existence of the threat, either as an incipient actual aggression or as a merely possible one, becomes verifiable. In such an event, the justification for pre-emption changes tense and transitions from one form of iterated modality to another: it is no longer a future possibility, a possible emergence as possibility, that is in question, but rather modals conjoined with past participles, that is, statements as to what would have happened had circumstances been different, or what would have occurred should certain opportunities have arisen. In this respect, Massumi attributes to the logic of pre-emption the claim that even if the threat itself—*a fortiori* the act defended against—fails to actualize, it remains the case that it *would* have done, *could* it have done so.[103] These tenuous and backward-looking justifications, which rely on modalities nested within modalities, Massumi refers to as instances of double conditionals. In turn, he describes them as a logic of affective legitimation.[104]

> The invasion was right because *in the past there was a future threat*. You cannot erase a "fact" like that. Just because the menace potential never became a clear and present danger doesn't mean that it wasn't there, all the more real for being nonexistent. The superlative futurity of unactualized threat feeds forward from the past, in a chicken run to the future past every intervening present. The threat *will have* been real for all eternity.[105]

It is natural to aim to understand in more precise terms the role of temporality in this idiosyncratic logic of pre-emption, and to relate it to the original questions of determinacy. First of all, a number of aspects of the framework involved in the philosophy of atemporality which clash with the pre-emptive interventions described above. In particular, consider how Russell's conception of a functional relationship between events $E_t = f(e_1, t_1, e_2, t_2, \ldots e_n, t_n, t)$ coheres with the modal characteristics of the systems described by Massumi. The $\{e_i\}$ are for Russell the events which occur at different times—present or future actualities—not so much double conditionals, the possibilities which could emerge. Regardless of the status of the function f, then, more is needed in order to provide an appropriate modal interpretation of sets of events such as $\{e_i\}$.

An expedient framing for this purpose is provided by Kazimierz Trzęsicki, particularly in his paper "Indeterministic Temporal Logic", which provides suitable formal tools for considering different types of logical relationships between propositions over time. Trzęsicki, working in the tradition of temporal logic established by Arthur Prior, distinguishes several different expressions of determinism and indeterminism and considers a range of methods to formalise them. The following are some straightforward examples.

[Determinism] If at time t α is true, then

> [Pre-Determinism] at any time $t_1 < t$, α *was* true at t, and
> [Post-Determinism.] at any time $t_1 > t$, α *will be* true at t.[106]

Immediately, there are suggestions of the distinction between two senses of these times—times indexing events and times at which events are determined—or, here, those at which the truth-values of the propositions describing these events are determined. There are even vestiges of the distinction between $\{t_i\}$ and $\{\tau_i\}$ at work in Esposito's notion of revisability. How can these theses (or their symbolic counterparts), which provide for relativizations of truth values to time, assist in assessing the modal character of Massumi's work?

In his first step towards constructing a logical system applicable to temporal determinacy, Trzęsicki introduces a triple $\langle T, <, V\rangle$ consisting of a set of moments T, binary ordering relation $<$ (allowing events to occur before or after one another), and a possible world V which consists of a set of propositions (atomic propositions) true at a time T.[107] This triple is used to provide definitions of various temporal operators, following the work of Arthur Prior. These operators, applied to propositions, formalise the prefixes which convey modal or temporal features—whether something will be the case, or couldn't have been the case, for instance. These operators render propositions true or false depending on the content of $\langle T, <, V\rangle$, t and can heuristically be taken as expressing formally modal statements about the future.[108] In this simple framing, Trzęsicki notes that the theses [PRE-DET] and [POST-DET] are valid independent of t. However, he then supplements this triple with a further important relation, which is decisive for the impact his presentation has on questions of modality and temporality. $\langle T, <, V\rangle$ becomes a quadruple $\langle T, <, \lhd, V\rangle$ with the inclusion of $\lhd$. Trzęsicki defines this $\lhd$ as a binary accessibility relation in the following way: Where $\langle T, <\rangle \times I$ is a class of times, $\lhd \langle\langle T, <\rangle \times I\rangle \times \langle\langle T, <\rangle \times I\rangle$.[109] I is a set which enables the addition of a notion of possibility to the description of a system, which Trzęsicki envisages as a branching structure.[110] The Cartesian product $\langle T, <\rangle \times I$ encompasses the totality of ordered moments each with many incompossible occurrences on different branches.[111] Most importantly, the relation of accessibility $\lhd$ restricts the times, that is, the members of a subset of $\langle T, <\rangle \times I$, in respect of which propositions can admit of truth or falsity. Trzęsicki goes on to define several temporal operators which modify those of Prior, taking account of this relation, which makes truth functionality of propositions about events conditional on the *accessibility* of events to other events or, more formally, on the accessibility of moments of time to other moments of time based on their order and relation to branching structure. As will shortly be shown, this relation of accessibility relativizes the determinacy of propositions' truth values to particular moments in time, and enables the logic of Massumi's *Ontopower* to be illustrated more precisely.

$$(G)$$

$\langle T, <, \lhd, V_i\rangle, t \vDash G\varphi$ iff for any $t_1 \in T, j \in I$:
if $(t, i) \lhd (t, j)$ and $t < t_1$, then $\langle T, <, \lhd, V_j\rangle, t_1 \vDash \varphi$.

$$(F)$$

$\langle T, <, \lhd, V_i\rangle, t \vDash F\varphi$ iff there is $t_1 \in T, t < t_1$ such that:
for any $j \in I$: if $(t, i) \lhd (t, j)$, then $\langle T, <, \lhd, V_j\rangle, t_1 \vDash \varphi$.

$$(H)$$

$\langle T, <, \lhd, V_i\rangle, t \vDash H\varphi$ iff for any $t_1 \in T, j \in I$:
if $(t, i) \lhd (t, j)$ and $t_1 < t$, then $\langle T, <, \lhd, V_j\rangle, t_1 \vDash \varphi$.

$$(P)$$

$\langle T, <, \lhd, V_i\rangle, t \vDash P\varphi$ iff there is $t_1 \in T$, $t_1 < t$ such that:
for any $j \in I$: if $(t, j) \lhd (t, i)$, then $\langle T, <, \lhd, V_j\rangle, t_1 \vDash \varphi$.[112]

Now, meeting the standard to qualify as a temporal operator applying to a proposition at a given time (or moment) requires specifying not only the relation of this time (or moment) to others through the relation <, but also consideration of which combinations of moments and branches are accessible at this time: $F\varphi$ holds only if φ is satisfied in any accessible possible world at the equivalent moment.[113] This limitation, among many other things, stops the incompossible or mutually incompatible evolutions which take place along different branches but whose times nonetheless stand in the relevant relations captured by < from being said to qualify as relevant determinations of propositions which concern features of each-other's branches which are not actually common. Without this limitation, in the context of any theory which posits a branching structure, these definitions would be opened up to absurdity and a range of counter-examples. Such evolutions would have little to say of a system's evolution over time: two paths which were to bifurcate at the outcome of a trial with contrary outcomes, such as a macroscopic coin flip or the measurement of the spin of an electron, could be answerable to the same set of truth-conditions in order to define the applicable temporal operators, despite their manifestly divergent trajectories.[114]

How, then, do *modal* operators fit into this picture? Like the temporal operators G to P, they are redefined from their more traditional instantiations in Priorean temporal logic.[115]

$$(\Box)$$

$\langle T, <, \lhd, V_i \rangle, t \vDash \Box\varphi$ iff
for any j : if $(t, i) \lhd (t, j)$, then $\langle T, <, \lhd, V_j \rangle, t \vDash \varphi$

$$(\Diamond)$$

$\langle T, <, \lhd, V_i \rangle, t \vDash \Diamond\varphi$ iff
there is j : if $(t, i) \lhd (t, j)$, then $\langle T, <, \lhd, V_j \rangle, t \vDash \varphi$

The thrust of these definitions is that the hypothesis of a branching structure, a structure which Massumi envisages in his own conceptualization of modality, enables the notions of necessity and possibility to be anchored in comparisons of truth-values across branches. One can assert something will be necessarily if and only if not only future moments make it so, but all moments in all accessible branches do so.[116] Trzęsicki's definitions and redefinitions allow formal equivalents of theses which pertain to temporality and determinacy to be expressed and evaluated and, most importantly, brought to bear on the kinds of systems Massumi's analysis concerns. The two theses which formalise the pre-deterministic and post-deterministic rubrics presented above have the following formal equivalents:

$$\Box\varphi \rightarrow HF\varphi$$

$$\Box\varphi \rightarrow GP\varphi$$

These two statements formalise pre-determinism and post-determinism in the following ways. As respects the former, the necessity of φ, taken as the truth of φ in all times and paths accessible from a given world with its concomitant set of time-relations, entails (i) that it has always been the case that at some time it would be the case that φ; and (ii) that it will always be the case that at some time it was the case that φ—both subject to the accessibility relations specified in (G) - (P), such that propositions truth-functional only on those moments in branches which are *accessible* from the time-point in question qualify or disqualify operators as applicable or not. The contention of Trzęsicki, however, is that both of these theses fail for the system, in the sense that they are not theorems. Simple counter-examples can be found rebuffing either, he shows. His proof involves considering two separate branching evolutions, one in which the relevant assignment of truth-values makes φ true and one in which φ is false. These branches begin separate and then converge, in the sense that the accessibility relation between the product of T and I initially fails to hold, but holds at a later time.[117] In such a scenario, $\Box\varphi$ holds as defined, since relative to a branch (and therefore subject to the relevant accessibility relation) one has that $\langle T, <, \lhd, V_j \rangle, t \models \varphi$. However, $GP\varphi$ fails: the definition of P requires that there be an earlier time at which, for *any* member of I from which the branch is accessible, φ holds. This is violated for Trzęsicki's assignment.[118]

It might be thought uncharitable of Trzęsicki's "Indeterministic Temporal Logic" to hang the metaphysics of post-determinism by the noose of an earliest time in the system being considered, as he does by defining post-determinism in such a manner as to require a previous time at which the proposition in question also holds. The finitude of past or future time is a source of unyielding controversy—and notoriously the subject of Kant's first Antinomy of Pure Reason in the Transcendental Dialectic,[119] in which both it and its negation yield contradictions—and thus hardly a truism to be used for the purposes of a refutation. There is, however, a more general point at stake in his argument against these two interpretations of determinism. The prospect of alternative histories leading up to a given confluence of circumstances in which propositions do or do not hold, or conversely the bifurcation of a history into competing futures, presents clear challenges for pre- and post-determinism, whether or not $\Box\varphi \rightarrow GP\varphi$ and $\Box\varphi \rightarrow HF\varphi$ are ultimately answerable to the framework and semantics Trzęsicki is wont to submit with the quadruple $\langle T, <, \lhd, V_i \rangle$. As respects the former thesis, branches diverging can make a proposition which is true in a subset of branches false in others accessible from the divergence point. As respects the latter, branches converging can make a proposition which is true in a subset of branches false in others accessible from the convergence point. If the values of the variables subject to a deterministic law or rule are plural for a given time, by virtue of each occupying a distinct branch in a ceaselessly ramifying arborescent or rhizomatic structure formed by the world's evolution, it is false to claim that for any one of these values that it did obtain or will obtain at other times. The branching structure need not introduce contingency into the system; branching is no reflection of randomness or stochasticity, for such structures are fundamental to deterministic theories in many cases, such as in the quantum domain.[120]

What, then, is the relation between the negation of these formalised theses and the temporality of determinacy, and how do these elucidate the approach of Massumi? To start with, quite clearly, pre-determinism and post-determinism conjointly hold propositions to be determined as true or false at all times before or after the moment at which the proposition holds. It is then trivial to infer from the assumption that the proposition is determined at this time that it is determined at all times. There is a natural alignment between this vantage point and the TVT, insofar as variation over time of truth-values is in both cases considered to confer temporal determinacy, albeit in the former case in the manner of a series of moments at which propositions are or are not true in contrast to transformations of occurrences of time-variables in the latter. Moreover, this approach fills some of the lacunae in the Russellian functional presentation by providing a framework in which logical relationships can hold between modal propositions. In this framework, temporal determinacy no longer involves the question as to functional relations between actual and determinate events, but extends it to possible or necessary events, whether or not interpreted as inhabitants of branches. If the core features of this logic accurately represent Massumi's position, if these features mandate the failure of a conjunction of propositions regarding determinacy, and if in turn an equivalence obtains between this conjunction and the atemporality conjecture, *affirming the antecedents in this chain means that Massumi's work entails a commitment to the temporality of determinacy.*

Let us assess the impact of these logical tools on the subject matter of this section. Recall the logic of affective legitimation[121] which Massumi finds in the discourse of interventionism—the iterated modality or double conditional whereby threats' mere *potentiality for becoming possible* elicits pre-emption. Massumi's statements as to, for instance, the putative offensive capabilities of a given threat actor—in the case of the 2003 war, the notorious weapons of mass destruction attributed to the Iraqi regime—can be structured in the common form *Pa*. To avoid confusion with the latter temporal operator let $Pa = Qa$. How must this atomic formula be amended or supplemented in order to reflect tensed and modal features, using the temporal and modal operators available—statements as to future becomings-possible, or as to what would have happened could circumstances have transpired suitably, which require more subtle orchestration? The implications of the above conclusions in the context of temporal logic for Massumi's theory will then be transparent.

Consider, in this vein, a few distinctive examples of the discourse of pre-emption, and their closest formal equivalents in this system of temporal and modal operators.

(I) $\Diamond\Diamond Qa$: A threat actor's attack could be possible.
Example: The worst threats must be confronted before they even emerge.[122]

(II) $\Diamond P \Diamond FQa$: It could have been the case that a threat actor's attack would be possible.
Example: The enemy could have had an offensive capability, which could have been transferred to those who might have subsequently used them.[123]

(III) $\Diamond PQa \rightarrow \Box FQa$: If a threat actor's attack had been possible, necessarily it would have happened.
Example: If the threat does not in fact materialize, it always would have, on the assumption that it could have.[124]

(IV) $\Box F \Diamond PQa$: It is always the case that a threat actor's attack will have been possible.
Example: For all of eternity, the threat *will have* existed.[125]

What do these propositions show as to temporal determinacy, and the atemporality conjecture in its several forms? For convenience, these are presented again here:

> Under the mapping $t \rightarrow t - t_0$ solutions of dynamical equations map to solutions of dynamical equations. (Translation Invariance Thesis)
>
> For any law f, and any occurrence of any $t_l \in \{t_i\}$, the truth-value of f does not change with variation in t_l (Truth-Value Constancy Thesis)
>
> For any law f, f is permanent (Permanence Thesis)

For no law f does the time-variable t occur in f where $dg'' + bg' + cg + d = f$

and $g' = \frac{\partial g}{\partial t}$, $g'' = \frac{\partial^2 g}{\partial t^2}$. (Time-Variable Thesis)

Assisting with this question, Trzęsicki's toolkit enables the propositions that have been considered so far in this text, normally predicated on functional relations, to be adapted for the purposes of Massumi's modalities. As noted above, the notion of a function f must be modified before being brought to bear on these modalities. Events occurring in worlds which evolve through time are phenomena common to both approaches. But functional relations, however intuitive in the context of an isolable physical system, feature less prominently in the context of pre-emption and its corresponding logical syntax, which consist rather of propositions describing possibilities and necessities in a variety of temporal relations. So how can the guises of the atemporality conjecture be examined in light of Massumi's work? The most natural equivalents to mathematical or quantitative relationships—captured by functions in the ambit of the latter—are *logical* relationships between propositions: in this language, quadruples of the form $\langle T, <, \lhd, V_j \rangle$, t, *qua* representations of possible worlds, distribute truth-values to propositions describing these worlds. The valuations described by V are functions which assigns to each point within T a set of propositional letters which obtain at this point.[126] $V \in \langle T, <, \lhd, V_j \rangle$, t thus connects times to the truth values of propositions describing events where f connects times and events. Now, to the extent that a meaningful relation obtains between the truth functions of this system and the quantitative functions occurring in the guises of the atemporality conjecture, the time-independence of whatever quantitative function f might be brought to bear on the various states inhabiting the kinds of worlds described by Massumi requires quadruples of the form $\langle T, <, \lhd, V_j \rangle$ to be time-independent also. In a sense, logical relationships replace quantitative laws. Yet this requirement would appear to be violated on multiple counts.

1. Firstly, as for the TVT, notwithstanding the differences between its constraints on f and the lack of a natural application of the schema

$dg'' + bg' + cg + d = f$ to the case of modal statements or to Trzęsicki's quadruples (which can hardly be assimilated to constituents of differential equations), there is a recognizable discordance produced when comparing this schema to a framework in which possible worlds are functions of time, and the truth or falsity of propositions about the world (and therefore the totality of facts) is determined by relations of accessibility and distributions of truth values effected by $\langle T, <, \lhd, V_j \rangle$, *t* which explicitly feature occurrences of *t*. If differential equations which determine the world's evolution also determine which propositions are true or false about the world, the TVT could be maintained only if the time-dependency assignment of truth values were eliminable from Trzęsicki's quadruples. Clearly this is not viable—they belong to the essence of the framework.

2. Secondly, and relatedly, as for the TVC, which concerns the truth-value of laws over time, consider again the behaviour of $\langle T, <, \lhd, V_j \rangle$. Patently, the assignment of truth values to propositions for members of *T* will involve different truth values being assigned to given propositions for different members *T*, for all but the most trivial mappings. Accordingly, *n*-tuples $t_1, t_2, \ldots, t_n$ will be discoverable such that $V(t_1) = V(t_2) = \ldots = V(t_n)$ is violated with respect to at least one proposition which is functional on *V*. The evolution of the world picked out by $\langle T, <, \lhd, V_j \rangle$ *consists of* time-dependent assignments of truth values to propositions describing its components. To expunge time-dependency from it in the manner of the TVC would be as abortive as to deny the evolution of the world itself.
3. As to permanence, it is less clear that there a strong rebuttal to be found in this strain of temporal and modal logic. Occurrences of events express the impermanence of the states of the world and correlative shifts in the truth values of the propositions describing them. But whilst the permanence of a law *f* can be given an intuitive sense by being moored to the consistency or uniformity of the phenomena it describes, as to the permanence of the equivalents of laws *f* in the system under consideration here, viz. $\langle T, <, \lhd, V_j \rangle$ with its assignments of truth values over time, being as they are *constituted* in changes which happen, no such gloss is forthcoming. Perhaps the only sense in

which a notion of permanence is accommodated by this framework is a higher-order notion of time-dependency: not the time-dependency constituted in the changes of truth-values over time, or the specificity of assignments to the members of T, but instead the fact that, for a given assignment $V(t_k)$ this assignment is constant; whether evaluated at t_k or at any other member of T, $V(t_k)$ remains the same. This, however, would underwrite only a weak notion of permanence, foreign to the permanence of laws of nature which crop up in descriptions of physical systems.

4. The final aspect of the atemporality conjecture to be looked at is the Translation Invariance Thesis. By now it is clear that solutions of dynamical equations are the possible event sets subject to f. Accordingly, the Translation Invariance Thesis' logical counterpart are the propositions describing these events. These may or may not be deducible from $\langle T, <, \lhd, V_j\rangle, t$. For solutions to map to solutions through a translation $t \rightarrow t - t_0$, one might expect that propositions would preserve their truth values under $\langle T, <, \lhd, V_j\rangle, t - t_0$ compared with $\langle T, <, \lhd, V_j\rangle, t$, so that $\langle T, <, \lhd, V_j\rangle, t - t_0 \vDash \varphi$ if and only if $\langle T, <, \lhd, V_j\rangle, t \vDash \varphi$. Nevertheless, this, like so many other comparisons of forms of the conjecture to this temporal and modal logic which formalises the basic components of the discourse of pre-emption, does not obtain.

This, then, surmises the implications of Massumi's analysis of pre-emption. The usages of iterated modalities with the various logical forms that they take are largely inconsistent with the four versions of the atemporality conjecture, and with other perspectives native to the atemporality of determinacy, such as the theses of pre- and post-determinism. These conclusions are reinforced by the properties of the most plausible available syntactic formalisations of the logical structure of Massumi's research.

However, there is another, even more important implication of this discourse for the temporality of determinacy, namely the sense which Massumi gives, or attempts to give, to a seeming antinomy: future causation, which is to say causation, affecting or effecting present events, out of and by future events. A number of passages from Massumi's *Ontopower* best evidence this commitment. In his definition of pre-emptive logic(s),

Massumi claims that the future should be assigned as much of a reality as the present, and regarded as possessing effective causal powers no less than does the present in linear causal models.[127] In a paradoxical way, the present itself, Massumi says, arrives after itself, coming about at a temporal lag: it is in and through its effects that it appears in its full import and meaning.[128] The present is pregnant with, temporalized by, acted upon by, the future, he suggests.[129]

Several elements of these excerpts impinge upon future causation—the reality of futurity and, concomitantly, the denial of presentism; the signification of events taken to manifest a threat depending on events which have not yet occurred; the lag or, in Deleuzian vernacular, caesura[130] which protracts the determinacy of the present; the co-presence of different modes of time. In what way does the phenomenon of pre-emption give sense to these notions? A simple model of Massumi's line of thought here can be formulated by means of the notion of threat. At a given time, a possible threat exists, formulable in the iterated modal expression $\Diamond \Diamond Qa$. Propositions as to the threat actor's impending action (an attack) are at this time undecideable and do not possess a determinate truth value. After the anticipatory strike on the putative threat actor, though, this indeterminate threat becomes real. Vengeance exacted by the opponent or even an act of self-defence confirms their status as an adversary along with their pre-existing hostile intentions; suspicions about future possibilities are replaced by conspiratorial inferences from presently observable conflict and destruction to anterior machination and scheming. This confirmation shows that at the time of the initial evaluation of the possible threat, an attack was in fact imminent, a confirmation which converts the hazy potentiality expressed in the sentence $\Diamond \Diamond Qa$ to the more determinate form $PFQa$. It is in this sense that the cause of the purportedly pre-emptive act, namely the threat actor's incipient attack, is futural. The pre-emptive act responds to a threat at a given time, constituted in an incipient attack which does not exist at this time; only after pre-emption has taken place does $PFQa$ obtain.

The "sign-event," the meaning of the inscrutable intentions of the adversary and the manoeuvres they obscure or reveal, is brought about by a future cause in so much as this possible threat becomes actual *only after* the moment in which it is appraised by the pre-empting actor. Their present intentions accordingly arrive at a lag revealing themselves only after they have been converted into active combatants. The originating cause whose presence is never discernible is given only as a past event, appearing in relation to its after-effects; the counter-strike which the pre-empted adversary considers as a response, is the after-effect of an intention absent at the time it was anticipated. Pre-emption is justified by the retaliation it elicits.

It is important here to distinguish between two possible interpretations of Massumi's notion of future causality, both of which have underpinnings in his writing, but only one of which coheres with the thrust of the present investigation into the temporality of determinacy. According to one interpretation, causal relations hold between events whose order in time is reversed: e_k causes e_l where $k > l$ for times t_k, t_l. But whilst the waning relevance of causation in metaphysical discussions might dovetail with the desire to deflate the concept into that of the determination of events by each-other, and accordingly into symmetric functional relations which endow with no more sense the past "causing" the future than they would the reverse, this interpretation, which in fact concerns causal relations of the traditional kind, namely, those typically taken to be asymmetric, attributes these characteristics to precisely the opposite sequence of events. Yet this problematic implication is one which could be gleaned from the prior passages, particularly with Massumi's attack on the privilege accorded to "linear causality" acting from the past to the future and the reluctance to acknowledge the same "effective reality" in the inverse direction. There is, however, an alternative interpretation of Massumi's approach, immune from the shortcomings of this fully inverted causality and more consonant with the formal representations already developed in this and the previous chapter for describing the temporality of determinacy. Rather than problematic causal relations being attributed to events which defy the natural ordering relations of events themselves, Massumi's account can be more suitably described in an alternative way—resonating with the conclusions of Esposito—by appealing to the distinction between the order of *events* ($\{ t_i \}$), and the order in which events are *determined* ($\{ \tau_i \}$).

How does this distinction operate in the discourse of pre-emption? Consider the following sequence of four events postulated by Massumi:

1. Threat actor A contemplates threat actor B as a possible threat
2. In contemplating such a possible threat, A considers the truth-value of the proposition as to whether B will attack A

A launches a pre-emptive attack on B
B launches a counter-strike on A

Imagine the times at which these four events occur as the members of the set $\{t_1, t_2, t_3, t_4\}$. Suppose that $t_1 = t_2 < t_3 < t_4$. Now, just as in the context of Esposito's analysis of expectations (interpreted as statistical functions) where the set $\{ \tau_i \}$ consists of times at which different revisions of these functions are made, in this context at least two distinct such times must be distinguished. The truth-value of the proposition " will attack A " is, considered by A at t_2, not determinate. In the event-language promulgated by Russell, the future event which is B 's attack on A, which might be assumed to be functional on data given at t_1 or t_2, is not in fact determined. Modal-temporal propositions such as $\Diamond \Diamond Qa$ accordingly replace the usual functional relations in descriptions of this system:

> At time t_p there is a time t_q such that $t_q > t_p$ such that e_q occurs, where e_q is an event in which A attacks B (pre-emption)
>
> At time t_p there is a time t_r such that $t_r > t_q$ such that e_r occurs, where e_r is an event in which B attacks A (counter-strike)

Later, however, in the transition from t_2 to t_3, A anticipates the indeterminate threat posed by B with an offensive act. B 's response at t_4 consummates their belligerence. The two or more values of $\{ \tau_i \}$ include as a minimum the time $\tau(t_1)$ at which 's future attack is undecideable and the time $\tau(t_4)$ at which it is decided. Consequently, in this case, $\tau_1 = \tau(t_1) = \tau(t_2) = \tau(t_3)$, whilst $\tau_4 = \tau(t_4) \neq \tau_1$. This latter fact, $\tau_4 \neq \tau_1$, prises open the distinction between the order of occurrence of the events in this sequence and the order in which they are determined. Only at τ_4 and not at τ_1 is the fact of the matter settled as to whether B will strike A, in spite of the fact that the t of which the latter is a function is earlier than that of the former. The notion of future causality is not therefore to be misconstrued as a traditional causal relationship with the direction of propagation reversed, with the obvious risks of absurdity and unreason this would entail, but rather the time lag over which earlier events finally become determined.

How does this interpretation of future causality reconcile with the letter of Massumi's publications? To start with, it coheres more closely with at least two morcels of textual evidence cited above. First of these is the claim that the present comes at a lag, which appears, not so much to reverse the permissible direction of propagation of causal relations, but rather to distinguish some aspect of the present moment which becomes determinate at a time *after* the time of this moment. Second of all is the claim that a sign-event can be triggered by a "future cause," in the sense, that is, that the signification or character of the event in question becomes determinate through a process. This perspective also explains the notion of recursion invoked by Massumi in the context of that of threat: the present state, comprised of iterations of possibilities, precipitates an action on which the threat itself depends. The identification of the threat becomes a self-verifying thought. The order in which the events in question are determined thus defies the order in which they occur, and the determinacy of the threat posed by the putative adversary considered here must be considered no less temporal, no less processual than the events themselves. This presentation of Massumi's work therefore allows the analogy with other thinkers of the temporality of determinacy in the context of the social sciences to be consolidated.

Consensus Algorithms and Temporality

This section focuses on one final domain in the non-physical sciences with interesting implications for the temporality of determinacy, namely distributed consensus in digitalized finance. The increasing utilisation of cryptocurrencies and the development of a market for trading them must rank among recent years' most intensively debated innovations in financial technology. Now surfacing commonly in discussions of alternative asset classes, wild fluctuations in the prices of many of the better known cryptocurrencies feature in the mainstream financial press and attract the attention of analysts, with large institutions now dedicating material human and financial capital to investigating their disruptive potential.

Speculative discussions on their long-term role in the financial sector reference anonymity, decentralisation, and security as distinctive, if disputed, benefits of this. As at 2020, recent estimates have pegged the capitalization of the market for cryptocurrencies at over USD 300bn, with daily traded volumes sometimes exceeding USD 100bn, and the rapid growth of attendant derivatives markets.[131] There has been commensurate intensification of the scholarly debate over these innovations—with a particular focus on the Bitcoin cryptocurrency—both as to their technological underpinnings and as to their broader significance, with the establishment of a dedicated peer-reviewed journal[132] published by the University of Pittsburgh, and academic discussions spanning the furthest reaches of philosophy, with one commentator alluding to Satoshi Nakomoto's 2008 paper[133] as the likely subject of a future in which philosophers regard it as tantamount to a kind of Pre-Socratic fragment.[134] In a paper delivered to a meeting of *The New Centre for Research and Practice*, the same author advances the rather extraordinary claim that these developments suggest the need for re-evaluation of the notions of simultaneity and synchrony at stake in fields in modern physics, viz. relativistic dynamics.[135] Such extreme philosophical fanfare mimics the maniacal fixation which has gripped the cryptocurrency market at various times, such as late 2017, leading to an extraordinary, if short-lived, hypertrophy in the price of Bitcoin, before a correction.

What are the driving forces behind this explosion in attention devoted to cryptocurrencies, both as harbingers of philosophical thought, as functional means of exchange and as investable assets? Answering this question, a task which demands an investigation into the technology at work in this other important social and economic system, reveals another aspect of the temporality of determinacy. Before identifying this, though, a short introduction is appropriate to the distinctive features of many cryptocurrencies and, in particular, the attainment of *distributed consensus* which is necessary for their functioning, and taken by some researchers to be among their definitive properties.[136] Following this introduction, this section will demonstrate how distributed consensus relates back to the temporal determinacy of the events it concerns.

Cryptocurrencies, such as the now well-known Bitcoin cryptocurrency, are currencies which enable transactions to be validated and recorded by decentralised, public and global networks of participants (nodes).[137] Whereas electronic transactions are typically reliant on intermediation by financial institutions and payment processors who authenticate transactions, transfer funds and maintain the integrity of a more or less central ledger, cryptocurrencies permit similar functionality by distributing the corresponding tasks to the network's constituent nodes. A ledger of transactions is maintained without a central authority, and a variety of cryptographic techniques are employed in order to verify ownership of the funds exchanged over the network and protect against attempts at forgery, digital counterfeiting and other fraudulent manipulations. These ultimately provide the basis for its advocates' boasts of security, anonymity and disintermediation.

With the Bitcoin cryptocurrency in particular, the public record takes the form of a chain of groups or blocks of transactions commonly described as the blockchain. This public record is particularly integral to the viability of the network, for the following reason: A given party's possessing a certain amount in Bitcoin means little more than being able to demonstrate that, at some point in the history the blockchain constitutes, the claimed amount was paid to the party and that these coins have not been spent subsequently—a straightforward demonstration.[138] Validity requires the following conditions being met:

1. The coins were in fact created by previous transactions.
2. The coins have not already been consumed in some previous transaction.
3. The total value of the coins is conserved throughout the transaction.
4. The transaction is validated, by being signed by the owners of these coins.[139]

Like many of the fields in the non-physical sciences drawn upon in this chapter, the salient property of cryptocurrencies such as Bitcoin as far as its present objectives are concerned is the way in which it popularizes a solution to problems concerning the order of certain events, where such "events" are simply the transactions occurring across the network. Because of the fact just noted, namely that ownership of some quantity of Bitcoin, together with the right to spend them, is conditional on pointing to previous transactions in which an equivalent or greater quantity of coins were received by the hypothesised owner, temporal and ordinal considerations figure in the criteria for the very validity of a transaction. That is, were it not possible for the blockchain to instil *consensus* between the network's participants on the order in which events have occurred, consensus over who possessed however many coins, and accordingly the right to spend them, would be similarly lacking. Since payments made in the cryptocurrency could not function without this kind of consensus, problems which centre on the order of transactions reach into the very heart of the feasibility of the currency. Accordingly, such problems are addressed relatively extensively in the literature.

Among the clearest illustrations of the necessity of achieving consensus on the order of transactions to the viability of these currencies is the widely discussed double-spend attack thought experiment.[140] This thought experiment is consistently drawn upon to illustrate the core functions of the blockchain technology. Consider a situation in which a participant in the blockchain network decides to pay a certain amount to a merchant's address in exchange for some goods or services, such as access to downloadable proprietary software. Payment consists in this participant successfully signalling to all nodes over the network that she intends to transfer a given amount of funds to a given recipient. Once received, the signalled transaction is incorporated into a block of roughly simultaneous transactions and appended to the chain of blocks of previous transactions, said chain ensuring that it can always be readily verified that the participant had, at this point in the chain, the right to spend this amount of coins they did. Once this spending is added as a block, the recipient of the funds would naturally conclude that she had been paid: she now has a block she can point to which evidences that the funds were transferred to her, which constitutes ownership of funds. However, consider now a situation in which she spends the very same coins she sent to the recipient at another node—one that, perhaps, she controls (meaning she has not really spent anything at all). Such a transaction is, as before, broadcast widely across the network. In turn, define latency as a constraint on the network such that signals sent do not proliferate to all nodes simultaneously: formally, clocks at different nodes do not assign the same time-values to the same signals they receive (signals announcing transactions to be incorporated into the blockchain). In such a case, for a set of events $\{e_i\}$ and associated times $\{t_j\}$ there will be different assignments of the latter to the former, that is, different pairings of each to the other, from node to node. If latency is taken to be in effect on the network, it follows that nodes may not agree on which signals are sent first, and therefore on which transactions occur in which order. Given further that nodes often extend the block with the transactions that they first received notice of on the network,[141] it follows in turn that there could be a lack of consensus about whether or not the transfer of funds to the merchant was at all a legitimate occurrence. The rule which dictates that nodes incorporate transactions that they heard about first is therefore a relative rather than absolute rule. Due to the latency involved in running the network, one could easily envisage a situation in which a prior transaction for one node were a later transaction for another, and in which a double-spend were inadvertently extended by an unwitting participant (miner).[142] Consequently, it is possible to exploit the latency of the network, and the resulting lack of consensus on the order in which candidate transactions have occurred, to defraud a payee or merchant, feigning the purchase of a good or service and then supplanting payment by diverting the blockchain along a path which includes an incompatible transaction.

How do cryptocurrencies, such as the Bitcoin cryptocurrency, resolve these problems? Failure to do so would result in a deeply flawed protocol. In order to function, they must instil consensus between participants on the *order* of transactions, i.e. the order of events occurring on the network. In Nakomoto's terms, participants must agree on the history of transactions, on the order in which payments were made.[143]

The solution to this requirement is termed the proof-of-work chain: a computational proof of sorts of the order in which transactions have occurred. In order for any transaction signalled to a given node to become a block appended to the chain, a computational puzzle (hash computation) must be solved which on average requires a material quantity of CPU expenditure. In order for a candidate transaction to be included in the chain, then, a substantial amount of computer power—which in reality requires the involvement of the entire network—must be exerted, so it is highly improbable that a node commanded by the double-spending fraudster would be able to extend the block in which she pays herself from the bifurcation point of the chain. With near-1 probability, it will be another block which produces the chain-extending block. If, further, the merchant seeking to avoid being defrauded adopts the principle that only after several blocks are appended to the chain will they consider payment to have been fully and finally made, it becomes negligibly improbable that a prospective double-spender could divert the chain to her advantage, short of controlling an unfeasible amount of the global network's CPU.[144]

Nakomoto goes on to specify a condensed six-step protocol in order to summarise how this approach produces a distributed consensus:

(1) When new transactions occur, they are communicated to all nodes.
(2) Any node receiving such communications incorporates the new transactions into a block.
(3) Such a node attempts to solve a proof-of-work problem for this block.
(4) If a proof-of-work is found, it is communicated across the entire network, to all nodes.
(5) All transactions in a broadcast block must be valid and not already spent in order to be accepted.
(6) Having accepted this block as valid, nodes begin working on the next block. This block includes the hash of the previous, recently accepted block.[145]

The difficulty of the proof-of-work problem that needs to be solved allows any participant to verify how much parallel, that is, cumulative, CPU must have been exerted in order for the problem to have been solved in the time given. Only if all nodes cooperated in this endeavour would this have happened, dispelling with near certainty the possibility of fraud, according to Nakamoto.[146] Blockchain-based protocols thus use peer-to-peer, distributed system in order to prove the order of transactions.[147, 148] Competing sets of proleptic transactions are broadcast across the network, with certain of the nodes which receive them incorporating the transactions into a record which requires a difficult cryptographic puzzle to be solved. Other nodes can verify this solution and the block containing transactions with minimal CPU expenditure. The objectivity of the record of which transactions have occurred is constituted in the difficulty of the puzzle, and the disincentive this presents to rogue agents such as double-spend attackers.

The problems which cryptocurrencies must overcome in order to avoid double-spend related difficulties are by no means new, considered in the broader context of distributed systems, albeit the stature of the pecuniary implications are to a degree. Cipher block chaining as a cryptographic technique was the subject of NIST standards in the 1980s, long before the advent of peer-to-peer payment systems such as these. Achieving a consensus association of times to events and enabling the synchronisation of clocks at distinct nodes of a network is a problem which has been analysed extensively in distributed systems theory, with specific theoretical problems and models built with the explicit purpose of addressing them.[149] While different machines are running different clocks, there is no guarantee of synchronicity.[150] Each clock, conceived abstractly as a process deemed regular relative to durations it serves to measure, and thus providing units which serve to count intervals of time, possesses distinct internal dynamics and initial conditions, whether down to intrinsic differences in their component crystals or extrinsic variables such as heat and humidity. Two clocks with ticks of differing frequencies will drift relative to one another over time, with one counting more intervals than its counterpart; clocks with identical frequencies may nonetheless differ in their assignments of times to events up to a constant term, if they are incepted asynchronously. Moreover, once one considers not just clocks' tick rates and start times, but their assignment of times to events qua signals they receive, another source of disagreement presents itself: the time taken for the signal transmitting an event to reach nodes at different distances differs between nodes. Thus, even assuming these clock rates and start times to be ideally matched, latency is bound to produce discord over the order of events.

In response, various ideal and practical algorithms have been developed in order to transcend some of these limitations and achieve synchrony or, as a minimum, consensus on certain temporal properties, such as the order of payments made in sequence. Cristian's algorithm, a simple algorithm for synchronisation in low-latency environments, simply dictates time from a master clock in the form of a time server, with a small adjustment made in order to obviate the effects of the lag between the initial signal and its arrival. A request made at t_1 from a client to a time server which is returned at t_2 enables the difference $\frac{1}{2}(t_2 - t_1)$ to be used as a proxy for the round-trip time, which is then added to the time T dictated by the server to establish synchrony. Alternatively, the Berkeley algorithm involves a master time server or "leader" which actively prompts all "follower" nodes to return their view of the present time. The leader aggregates these views, filters them for outliers and then averages them, issuing commands to the clocks requiring individual adjustments to harmonize them all. This negates the effect of local drift which sees nodes' clocks ticking relatively faster or slower. Thirdly, and most germane to the present discussion, Leslie Lamport's research into and development of the now frequently referenced Paxos algorithm[151] considers ways of reaching consensus in distributed systems—systems consisting of agents communicating with one another—with various more realistic (i.e. inconvenient) properties: agents can operate at arbitrary speeds, and can fail and have to stop or restart; messages can take arbitrary amounts of time to reach their destination, and are subject to possible duplication and loss.[152] The essence of rudimentary models of the Paxos approach involve a distinction between a client, acceptor, proposer, learner and leader.[153] Clients send requests, perhaps to write a given file to the distributed system or to establish a consensus transaction history, to the system. Acceptors, grouped into quorums, are responsible for agreeing on which of a number of potentially inconsistent requests are to be accepted. Proposers propose these competing requests to acceptors, with a view to inculcating agreement, which is then relayed to learners. The leader is a privileged proposer whose recommendations are submitted to and agreed by quorums of acceptors, a privilege which can pass to new leaders if consensus fails to be reached. In implementations, these roles may not be exclusive, in the sense that agents in the system may fulfil more than one.[154] Modifications of the algorithm can be made to accommodate situations in which the model's assumptions are relaxed, e.g. when rogue agents circulating duplicitous messages populate the system alongside loyal ones, one example of which is the Byzantine Paxos approach, a reference to the eponymous fault tolerance problem.[155]

This protocol bears structural resemblances to the implementations which appear in the cryptographic framework used by digital currencies. As outlined above, these currencies deploy protocols which allow participants to distinguish valid from invalid transactions based on the order of events occurring in the chain, with valid payments defined as just those which expend cryptocurrency verifiably remitted to the payer in previous blocks and not expended since.[156] This protocol is, of course, a convention. There is nothing absolute about the way different cryptocurrencies are set up to run by developers; rather, there is an entire class of quite fluid protocols, or potential protocols, which could be used to instil consensus, each with different and often conflicting merits and demerits. Indeed, cryptocurrencies such as ethereum have even undergone hard forks,[157,158] points of bifurcation into permanently distinct branches where groups of participants who have reached consensus on the chain prior to the bifurcation point come to disagree on which chain after this point contains the legitimate events in the legitimate order, thus splitting consensus.

This conventionality is related to another important property which secures the connection to the temporality of determinacy. In the context of distributed systems, determination consists in attainment of consensus across nodes or agents interacting over the network. Here, the relevant form of consensus contemplates the order of events, taken as for instance the order in which transactions have occurred, with the usual $\{e_i\}$ consisting of payments made between agents with the $\{t_i\}$ setting their order. As has been illustrated, such a consensus is a prerequisite for the viability of the currency given the threat of digital counterfeiting.[159] Consider, then, what the nature of this consensus implies for the determinacy of these events.

First of all, could they be so determinate as to fulfil the definition of determinism introduced in earlier chapters? Depending on the precise details, nothing in principle precludes this eventuality. It would appear to depend on the status of the requests sent by clients and referred by proposers, or, in the case of applications of similar protocols in cryptocurrencies, on the *markets* in relation to which this currency functions as a means of exchange: who sells what to whom, when, and for what; capital flows in and out of the markets for the currency; regulatory and other restrictions on trade, etc. Satisfying the Russellian definition would require a function of the form $f(e_1, t_1, e_2, t_2, \ldots e_n, t_n, t)$ relating exchanges of cryptocurrency at their respective times, each member of $\{e_i\}$ being in turn functional on these and other variables, whether the subjective utility which motivates a consumer's preferences, speculative injections of capital into the crypto market, trading operations at exchanges and so on. Thus, whilst the decisions of agents such as clients and leaders might not be describable in this form for any remotely tractable value of n (that is, where $\{e_i\} = \{e_1, e_2, \ldots, e_n\}$)—excepting perhaps certain forms of functionalism in the philosophy of mind—scenarios can be *envisaged* in which the actors on the network send, propose, accept, and learn requests according to simply defined deterministic rules. The applicability of functions such as this thus depends somewhat on whether one wishes to consider theoretical models populated by automata, or rather the real-world, subject as it is to the vicissitudes of subjective decision making and complex social systems, viz., the systems of consumer preferences and capital flows.

However, second of all, and aside from the broader question as to the deterministic or indeterministic nature of these distributed systems, one cannot fail to notice the dynamic nature of consensus. Return to the sequence of cryptocurrency payments e_i. The determinacy of each payment requires legitimation by consensus. As discussed above, candidate transactions broadcast to subsets or quorums of participants are incorporated into proposals; these are temporally indexed in order to forestall digital counterfeiting and double-spend attacks. In the case of blockchain-based cryptocurrencies, though, this consensus is dynamic, because of the branching (if not quite arborescent) structure of the chain.[160] As before, this structure arises from the possibility of *competing* blocks found by different miners and communicated to distant nodes within sufficiently small intervals of time producing nascent bifurcations, and temporary discord as to the order of transactions. On what basis are the competing blocks included or excluded? They are included or excluded ultimately by no endogenous rule. It remains forever possible, albeit computationally unfeasible and exponentially *improbable* as more and more blocks are appended, that transactions can be *de facto* superceded by incompatible ones. As will now be demonstrated, it is for this reason that the chain's ineliminable potential for branching temporalizes the legitimation of transactions.

Recall that the relevant definition of "determinacy" in this context pertains to transactions becoming subject to consensus (as to which have occurred and in what order), attained progressively among participants. Consider a candidate transaction e_v, between two participants in the network P_1 and P_2, which takes place at a time t_v. In this instance, let P_1 be the purchaser of some goods priced at z units of cryptocurrency. Immediately prior to t_v, let P_1 possess a balance of no greater than z units. Consider further a candidate transaction e_w which takes place at a time t_w, which involves P_1 remitting a finite amount of currency to another party P_3, either identical to or otherwise in collusion with P_1. P_1 's balance means that this latter transaction is incompatible with e_v.[161] Intuitively, if $t_v < t_w$, only e_v is legitimate; if $t_w < t_v$, only e_w. The crucial question then concerns the nature of the binary relation $<$: what determines which ordered pairs are so related? Insofar as determinacy means consensus attained as to temporal order, the relation $<$ picks out which members of pairs occupy blocks earlier in the consensus chain than those occupied by their counterparts. Consider, by contrast, the relationship these bear to a temporal parameter τ largely independent of the operation of the distributed system, such as Coordinated Universal Time (UCT). Given that different branching networks can ascribe different relations $<$ to pairs drawn from $\{t_i\}$, and that these different networks might arise in the context of a double-spend, block-withholding attack, or other failure of consensus which could split the chain and orphan groups of blocks, it is possible to find τ_v, τ_w where $\tau_v \neq \tau_w$, such that $t_v < t_w$ for τ_v and yet $t_w < t_v$ for τ_w. Whereas in the latter case only e_w is legitimate, in the former, so is only e_v. The subterfuge *succeeds subject to* τ. Thus, the most relevant parameter for determining which are the legitimate candidate transactions $\{e_i\}$ on the network, namely the $\{t_i\}$ defined by the consensus sequence of blocks, is itself temporal, as long as different blocks are distinguishable and there exists the possibility of re-ordering the $\{t_i\}$.[162]

In this regard, these algorithms and cryptographic protocols invented in order to achieve consensus in distributed systems harbour an even more extreme temporalization of the dynamics of the system of transactions that they describe. The work of Esposito implied two features which, it was claimed, delineated her impact of her theory on the question of the temporality of determinacy. On the one hand, the determination of events became progressive, became a process. On the other hand, the order of events (indexed by $\{ t_i \}$) and the order in which they are determined (indexed by $\{ \tau_i \}$) are distinct. In the two sections which followed these claims, evidence was presented that these attributes hold also in the case of other fields in the non-physical sciences, such as the analysis of time series common in economics and finance, and in the logic of threat and pre-emption driving recent multi-polar conflicts. In the present case of distributed systems and the consensus algorithms deployed in them, though, it is not only these two conditions, relevant to the temporality of determinacy, which are fulfilled—because of the need for another time $\{ \tau_i \}$, whether UTC or another standard, which distinguishes different branches containing incompossible chains of transactions. Going beyond Esposito, there remains a distinct possibility: that the order of the events themselves, the $\{t_i\}$ associated with the $\{e_i\}$, vary. Thus, for different values of τ, different time-orders prevail, as a consequence of the possibility that at two different moments distinguished by an clock external to the distributed system two different chains assign different temporal coordinates to a given pair of events. The stability of these coordinates, the fact that there are blocks far enough back in the chain for their position to be taken as more or less fixed and highly unlikely to be revised, the fact they are used as a base for mining subsequent ones—all these features reflect a level of *relative* credence given to the convention elected by participants. Without the collective choice to regard blocks mined in the distant past (which is to say mined at a time $t_r \ll t_s$ where t_s is the current block) as permanently determinate there would be no basis for the fixity of past transactions. The order of the blocks and the relative distribution $\{t_i\}$ is thus relatively stable and strictly variable as would be, accordingly, the two-place relation $<$ between pairs of these times. In such a scenario, it would be misleading to regard τ as a simple function of t, monotonic or not, as previously might have been implied by writing $\tau = \tau(t)$ when considering the variation of the expectation functions for the values of different securities. In fact, in this event, different regimes of τ, whether driven by totemic changes in the distribution of CPU, changes in the central protocol of the system, or swings in consensus, reorder t.

Realistically, few if any species of the atemporality conjecture can be reconciled to this group of properties. The notion of permanence is thoroughly incongruous with the revisable blockchain sequence, even if any interference with blocks mined in the distant past (relative to real time) is parabolically challenging.[163] As for the others—the TVT for example—which discuss occurrences of the time variable t in dynamical equations, the closest equivalents which allow these hypotheses to be tested in the context of distributed systems and crypto-assets would relate to the algorithm. Dynamical laws may or may not exist conditioning the circuitry used to evaluate hash functions, the ambient temperature which affects the functionality of mining centres, the neurochemistry of participants exchanging currency over the network, or other variables which influence the nature and timing of the transactions. But inscribed into the algorithm which acts as the rulebook[164] for transacting the asset, such as the Bitcoin Core software, controlling the time-stamping, validation and communication of transactions leading to consensus, are references and connections to other blocks which validate or invalidate the transaction in question,[165] blocks and transactions which are themselves indexed by time. Such indexations and dependencies can hardly be removed in the simple way the atemporality theses would imply. These synchronisation protocols in distributed systems, concretized clearly in the emergence of digital currencies and the blockchain technology, thus provide another example of a system, evolving according to a rule, which firmly resists the blunt theoretical tools of the atemporality thesis and its sub-doctrines, and endows temporal determinacy with further, more radical, sense.

Conclusion

The analysis of time-dependency can be brought to bear quite naturally on social systems. Populated by agents and actors making decisions and forming expectations on the basis of others', these systems admit temporal and modal structures which challenge the conceptual arsenal of stationarity, atemporality and linearity. Whether considered ideally rational utility maximizers, victims of cognitive and emotional biases, or exhibiting whatever other traits theories ascribe to them, the inter-dependencies of participants in a social system generate counter-intuitive sequence effects, making them challenging to describe and predict. Outcomes depend on the context and environment in which decisions are made; the way in which events are ordered influences the evolution of systems they inhabit; the nature and meaning of attributes described by theories shift and metamorphose over time.

This is not to overstate the similarities between these fields. Whilst all consider the ways in which assumptions about time break down for complex systems of agents and actors, there are important differences between them. The processes described by Esposito are revisions of expectations about markets and prices over time, whereas Massumi's are adversaries' evolving tactics in a conflict. Abbott considers a very broad range of sequence effects within social and political reality, such as the causes of criminal behaviour and inequality. Distributed consensus protocols subject the order in which financial transactions take place to a processual algorithm. These different contexts involve different dynamics. All the associated processes, however, have in common a distinction between the order of events and the process which brings about this order. The dynamics of the system establish the order.

All of these theories consider the way in which agents interact, whether the theories are concerned with nodes transacting across a distributed system or participants in derivative markets. One might ask: from where arise these idiosyncratic properties, namely the appearance of dynamics which accommodate the temporality of determinacy and which refuse atemporal descriptions, anchored in such characteristics as permanence and time-independence? One immediate suggestion presents itself. It may be thought that these properties reflect not so much the dynamics of the system at a fundamental level, but rather the nature of the description of the systems under consideration. By considering systems populated by agents and nodes, one brackets the forces which dictate the evolution of the system at a level of analysis which is finer, lower or deeper (in a sense to be specified). Most attention is thus given to macroscopic (rather than underlying, microscopic) behaviour. Nodes and traders are themselves composed of highly complex subsystems the likes of which might be studied by neuroscience, chemistry and a number of other disciplines. But it follows no more from the fact that phenomenological characteristics hold of a composite system that they do of its component sub-systems too than it is a truism that emergent properties do not exist. One cannot legitimately infer from the failure of atemporality theses at the level of a system that all sub-systems refute these theses also. At least, this is the sort of inference which would require some justification and defence. It might therefore be concluded that the applicability of conclusions about temporal determinacy is limited in the context of such systems—limited to the *macroscopic totality* of the system under consideration. As to whether such conclusions apply also to the system's internal components, one would be forced to pass over the question in silence.

An attack vector such as this is particularly virulent in the domain of time-series analysis, which provides a useful illustration. This discipline makes far less enquiry into the driving forces behind the appearance of data in series than it does into the panoply of formal techniques to describe these data. Predictions and explanations are then couched in terms of sufficiently close-fitting statistical models. There is no sense in which the values $\{y_i\}$ at times $\{t_i\}$ are terminal values, which preclude an underlying set of values which could be explanatory of these or obey simpler functional relationships with fewer variables and smaller error terms. Such a set $\{z_j\}$ could have the property that, for any $y_k \in \{y_i\}$, a related $z_l \in \{z_j\}$ would exist and that, for whatever statistical relation holds between the $\{y_i\}$, some more simple relation holds between the $\{z_j\}$ eliminating the residuals w_t and replacing them with, perhaps, a straightforward constant term C which does not vary with time. The possibility of such a reductive mapping may militate against reading the nature of determinacy off theories which have a statistical validity. Such is precisely the line of thought which, for one, Ray Brassier adopts in response to Esposito's innovations,[166] by postulating dynamics, such as changes in the mental states of market participants, which may not display the same characteristics as the overlying formalism.

As to the possibility of a reduction being performed to simpler combinations of variables, simplicity is a notoriously elusive criterion for distinguishing more from less fundamental theories. It is not, however, unprecedented—neither as an aesthetic criterion for scientific laws, nor as the foundation for the development of theories. In the development of the theory of special relativity, for instance, the realization of the *conventional,* rather than absolute, nature of simultaneity led to Poincaré's and, later, Einstein's insight that conventions used were to be those which reduced the fundamental dynamical equations to their simplest possible form: the order of events taking place, or relations of simultaneity which hold between them, must be defined so that the laws to which they are subject take the simplest possible form; time is a parameter which is chosen or selected by the observer, in such a way as to simplify the equations of motion describing them to the greatest possible extent.[167, 168] However, it remains rather unclear how to rigourise this criterion in such a way as could ground the reduction of $\{y_i\}$ to $\{z_j\}$, despite scholars' proclivity for positing underlying laws with more appealing structures than cumbersome superficial descriptions. If objections to the characteristics apparent in time-series models (which could be taken as manifestations of temporal determinacy) rely on criteria for reducing these models to forms in which these characteristics no longer appear, these criteria should be to some degree specifiable. But there is little present prospect of actually substituting for the range of descriptive techniques employed in the analysis of time-series simpler time-independent descriptions, that is, developing a fundamental theory of markets, with the unaccountably vast degrees of freedom that they possess. Thus whilst this line of objection tries to reduce the scope of talk about time-dependency to non-fundamental descriptions, it does not *specify* relations which reduce time-series relations to simpler, more fundamental relations with a different temporal status. This is surely a defect of any objection to temporal determinacy based on this approach.

But there is a more disabling problem for this attempt to weaken manifestations of temporal determinacy in the non-physical sciences. In order to license an inference from either the existence or the possibility of simpler, more fundamental law-like descriptions to the denial that their properties are general, it must be assumed that a necessary condition for such generality lies in the fundamental nature of these descriptions. Only if assurances exist that these are ground-up, rather than of statistical validity, are their structural features to be taken seriously. Yet this criterion fails for countless theories and laws whose structural features are perfectly legitimately objects of analysis in the philosophy of science:

1. Nobody familiar with the developments in quantum mechanics in the early 20th Century would hold Newtonian physics to be a completely fundamental theory in the sense that its subject systems could not be further reduced to sub-systems which obey different laws, approximately generative of Newtonian ones.
2. Recovering the classical limit as quantum parameters (such as Planck's constant) are scaled is, rightly or wrongly, frequently spoken of as a constraint on any permissible formulation of quantum mechanics.[169] As D. Z. Albert remarks in his *Quantum Mechanics and Experience*, to the extent that classical mechanics provides a very valid description of larger systems, it ought to be possible to approximately reproduce these descriptions within any quantum theory subject to certain constraints.[170]
3. Statistical mechanics, developed as an alternative model of the behaviour of many-particle systems such as gases, puts aside microscopic changes in the motions of individual particles. Kinetic theory, which applies classical mechanics to these particles, is a more fundamental approach providing an explanation of the system's behaviour in terms of the dynamics of its component sub-systems. Yet doing so requires manipulating unfeasibly large numbers of equations, hence the greater expediency of deploying statistical relations between simple observables such as pressure, temperature and volume.

What is the significance of these examples? Despite that they have to do with laws which are in no way considered to be fundamental, they amount to prominent theories which facilitate and contribute evidence to debates about their metaphysical implications. Newton's laws, such as the law of gravitation, are invoked in discussions of action at a distance, as evidence that forces can propagate instantaneously (a problematic implication not only in the eyes of Newton's detractors, but by his own admission).[171] Statistical mechanics and kinetic theory are frequently contrasted in discussions of the thermodynamic arrow of time: whereas the dynamical equations of kinetic theory are such that temporal evolution in either of two directions is physically possible, statistical mechanics picks out a preferred direction in time (consider how often one observes shattered ceramic spontaneously reassembling itself into a cup, an elderly person ageing into their youth, or convergent waves in a pond meeting at a point and ejecting a pebble from the surface of the water, despite the consistency of both processes with "fundamental" dynamics—a fact that in no way disqualifies, but rather intensely problematizes, the observation of one and not the other process). This clearly violates the condition that only fundamental descriptions are to be used as a basis to draw conclusions about time-dependency or other aspects of the architecture of laws. It is accordingly dubious as to whether fundamentality of laws can reasonably be taken as a necessary condition for supporting conclusions about time-dependency.

For further illustration, return to the example of time series analysis. It could well be thought that the foregoing analysis of time series methods fails to prove the absence of atemporal functional relations between data. The undeniable kernel of this objection is the truism that it in no way follows from the failure to find a function *f* meeting the criteria for atemporal determinacy that no such function could be found; it is the existence, rather than the discoverability, of the functions which is relevant.[172] Nor, therefore, does it follow that whatever deterministic properties entail and presuppose the existence of *f* fail to obtain for these systems. After all, systems which are commonly regarded as conforming to such functions in areas of physical science—classical Newtonian systems, for instance—frequently find observation data which do not exactly fulfil models, but rather do so with a tolerable level of precision. Tests of Newton's laws of gravitation might hold only if minute perturbations effected by distant objects (such as planets on the other side of the universe) are disregarded, on the basis that the law of gravitation falls off with the square of the distance to the gravitating object, or if certain simplifications are made as to the shape of the gravitating bodies. The notion of statistical significance is in this sense as relevant to classical systems in the physical sciences as it is to the subject disciplines in the foundations of finance and economics, providing grounds for separating verified or falsified empirical hypotheses from inconclusive ones. So why ought the prevalence of such approximations in the context of time series analysis be taken to rule out other laws, which do not in fact display the same temporal structure?

As much as this reasoning is correct, it misconstrues the purpose of the argument put forward in this section. It is not so much that the fact that the functions that have been found are approximate licenses the conclusion that no exact function exists—an unmistakeable *non-sequitur*. The argument is rather that the existence of exact functions is not assumed to be part of the disciplines being studied. One does not need to hold that a particular discipline has arrived at knowledge of all fundamental laws of its systems in order for it to be used to cast light on conceptual questions. Thus, the argument of this section in no way rests upon the completeness or finality of the theories on which it draws. Accordingly, it is no impediment to an analysis of the time-dependency latent in scientific laws that they may be reducible to more fundamental ones. Such an analysis is especially germane in the non-physical sciences, where the relevant laws explicitly contemplate highly complex systems populated by composite entities, such as governments, digital networks and decision makers. Nonetheless, it is quite natural to ask whether this time-dependency—whether constituted in the failure of various of the atemporality conjectures, the processual nature of determination in certain social systems, or the distinction between the order of events and the order of their determination, as in Esposito and Massumi—holds all the way down. Are the most fundamental theories, assuming they exist, also time-dependent? What inferences can be drawn about them based on what we know of their macrophysical dependents?

As discussed in the introduction to this work, some scholars have insisted for reasons of philosophical principle (rather than for the reason that bottom-up developments in the sciences compel it) that fundamental theories would indeed have to share these characteristics. C. S. Peirce declares as much in his tract "The Architecture of Physical Theories," speaking of a natural history of laws of nature. To declare certain such laws inexplicable, unassailable, infallible, and refuse to justify or ground these is hardly consonant with the spirit of scientific endeavour; rather, such laws and the uniformities they account for surely require explanation—an explanation which, for Peirce, lies in its *evolution*.[173] *Any* uniformity, whether grounding or grounded, needs to be justified. Accordingly, the present world's obedience is something which has progressed through time, from an original state of fortuity dominated by random perturbations into a state of thoroughgoing determination under rules.[174] As will be analysed extensively in the next chapter of this present work, Lee

Smolin's research (from *Time Reborn* to *Einstein's Unfinished Revolution* to *Life of the Cosmos*) considers a similar evolutionary process governing laws, encouraging metaphysics and science to jettison its support for timeless and eternal laws of nature,[175] and speaks of a process of selection governing laws over time,[176] analogous to the more familiar process of natural selection.[177]

For the question at hand, these provocative questions ultimately stimulate more conjecture than they do constructive discussion. The analysis presented in this chapter aims to build a plausible case for temporal determinacy based on evidence and example rather than *a priori* principle. It draws upon as broad a base of evidence as possible: the foundations of social science, the foundations of finance, the logic and temporality of political and military theory, time-series analysis and protocols in distributed systems. The fact that the theories involved may one day be regarded as wrong or incomplete, or mapped to deeper theories which undergird them, does not delegitimize this approach. The time-dependent structures delimited in this chapter need not belong to theories which are final in order to refute the atemporality conjectures nor to give meaning to the time-dependency of laws. For, recognising that theories are fallible, the best available candidates for successful ones are nonetheless legitimate objects for philosophical analysis. Otherwise, the falsifiability of a hypothesis such as this would amount to a falsification of it, and any analysis of physical laws would be suspended until they are replaced, in the infinite long-run, by a theory which would be perfectly complete.

Notes

1. Abbott (2001), p. 37.
2. Abbott (2001), p. 38.
3. Abbott (2016), p. 238.
4. Abbott (2001), p. 39.
5. Abbott (2016), pp. xiv, 233–252.
6. Abbott (2016), pp. 180–181.
7. Abbott (2016), pp. ix–xi.
8. Nietzsche (1967), p. 550.

9. Abbott (2001), p. 38.
10. Abbott (2001), p. 47.
11. Abbott (2001), p. 51.
12. Abbott (2001), pp. 56–57.
13. Occasionally these are even mapped onto specific philosophical debates; in Abbott's essay "The Idea of Outcome" he considers the assumptions of empirical sociology and microeconomics in terms of the philosophy of time of McTaggart, Bergson and Aristotle. See Abbott (2016), pp. 179–184.
14. Robinson (1981), p.
15. Robinson (1981), p. 89.
16. Esposito (2011), p. 9.
17. Esposito (2011), p. 6.
18. Keynes (1936), p. 9.
19. Keynes (1936), p. 25.
20. Malik (2014), pp. 629–812.
21. Soros (1987), p. 26.
22. Esposito (2011), p. 40.
23. Esposito (2011), p. 105.
24. Esposito (2011), p. 54.
25. Esposito (2011), p. 11.
26. Esposito (2011), p. 22.
27. Husbands (2020a), p. 117.
28. This terminology Esposito takes over from Niklas Luhmann, particularly Luhmann (1981), pp. 126–50.
29. Esposito (2011), p. 23.
30. Brown (2005), p. 19.
31. Barbour (1989).
32. Esposito (2011), p. 23.
33. Newton (1999).
34. Esposito (2011), p. 22.
35. Fama (1970).
36. Fama (1970), p. 414.
37. Soros (1987), p. 2.
38. See also on this notion of reflexivity the work of Malik (2014), pp. 717–718.
39. Soros (1987), p. 29.
40. Esposito (2011), p. 136.

41. Brassier et al. (2017), p. 95.
42. Brassier et al. (2017). It should be noted that Brassier's account addresses primarily Suhail Malik's essay "The Ontology of Finance" in the 2014 edition of *Collapse*. In this respect, Brassier addresses Esposito's argument secondarily, that is, to the extent that Malik's research relies heavily on hers.
43. Esposito (2011), pp. 23, 27–28.
44. Brassier et al. (2017).
45. Esposito (2011), p. 136.
46. Jon Roffe establishes a similar distinction in his own research in the theory of finance *Abstract Market Theory*. See Roffe (2015), p. 4.
47. Esposito (2011), p. 136.
48. It is worth observing that whilst this case compares the expected value of the future spot price with the spot price at the reference time, a related question arises in the context of discussions which compare the former to the price of a futures contract on the same underlying asset. Futures are standardized, exchange-traded contracts which oblige two parties to exchange a good on a specified date for a specified price, though rarely involving the delivery of the reference asset and often transacted for speculative purposes. These discussions consider why the expected value of the spot price three months later should differ from the traded futures price. They attempt to account for this difference in terms of determinate and evaluable circumstantial factors: storage costs, dividends, convenience yields, interest rates, risk reduction benefits, correlation with other markets and other variables, all of which make the temporal interval between the formation of the initial expected value and the future three months later a source of inequivalence. There is, however, a residual question as to whether these variables could ever provide a complete account of this inequivalence. To the extent that these differ in ways unaccountable by determinate factors, this is because neither the value of the asset nor the worth is not determined, but only modelled output, simulated worth, or subjective instinct, conferring a statistical or judgmental rather than an objective meaning. See e.g. Hull (2002), pp. 56–63.
49. Even today, financial commentators on asset classes which are frequently described as bubbles refer to the questions over this description as moot until after the bubble is (or is not) burst; see for a recent exam-

ple UBS Global Real Estate Bubble Index, UBS Chief Investment Office (2020).
50. Esposito (2011), pp. 27–28.
51. Hull (2002), p. 246.
52. Esposito (2011), p. 139.
53. Esposito (2011), p. 9.
54. Esposito (2011), p. 134.
55. Malik (2014), pp. 629–636.
56. Esposito (2011), p. 105.
57. A similar distinction to Esposito's, the distinction between market pricing and value, is analysed at length in Roffe (2015), and Mackenzie (2006).
58. These confer the right but not the obligation to purchase a security at an agreed strike price at an agreed time, or within an agreed time-window. See Hull (2002).
59. It is interesting and surprising to observe that in spite of their almost entirely antithetical line of thinking on this matter, Lindsay and Margenau intimate the existence of a similar distinction between different temporal parameters at work in systems and their descriptions, both with their modulating criteria for the time-independence of a law, and with their limitation of the scope of their claims about the time-independence of laws to closed causal systems, paralleling the notion of endogeneity which appears in the rather different context of Esposito's work. See Lindsay and Margenau (1963) and Margenau (1977).
60. Russell (1918), p. 199.
61. This notation follows that of Anderson and Zalta (2004).
62. Esposito (2018), pp. 219–220.
63. Esposito (2011), pp. 24–25.
64. Johnson (2016), p. 189.
65. Roffe (2015), p. 150.
66. Roffe goes so far as to declare that belief in the reliability of pricing models constitutes a kind of mental sickness, and ascribes an absolute uncertainty, unknowability and unpredictability to future states. Roffe (2015), p. 14.
67. Malik (2014), p. 715.
68. Malik (2014), p. 718.

69. See Ayache (2010) and Ayache (2015). The latter (p. 137) features his claim that the derivatives market could be regarded as a form of Brownian motion.
70. See Mackenzie (2006).
71. Hamilton (1994).
72. Campbell, Lo and MacKinlay (1997), pp. 8–9.
73. Brock and De Lima (1996).
74. Hamilton (1994), p. 25.
75. This, provided a suitable definition of the matrices, coheres with Abbott's framing of the role of the GLM which, as recounted earlier in this chapter, can be expressed in the form $\boldsymbol{X}_t = \boldsymbol{X}_{t-1}\boldsymbol{B} + \boldsymbol{U}$.
76. Hamilton (1994), p. 45.
77. Chatfield and Xing (2019), p. 17.
78. Chatfield and Xing (2019), p. 43.
79. Hamilton (1994) and Chatfield and Xing (2019), p. 16.
80. Chatfield and Xing (2019), p. 41.
81. Chatfield and Xing (2019), p. 42.
82. Chatfield and Xing (2019), p. 9 allude to this relationship as one of partial determination of the system's behaviour given initial (past) conditions—as well as once again disinterring the problematic relationship between determinism and predictability. Deterministic time series are those which can be predicted exactly, stochastic time series are those for which the history of the system determines its future only to a partial degree, meaning future outcomes are best described by a probability distribution.
83. Chatfield and Xing (2019), p. 42.
84. Chatfield and Xing (2019), p. 45.
85. Berkovitz et al. (2006), pp. 668–671.
86. Grimmett and Stirzaker (2001), p. 47.
87. More formally, Grimmett and Stirzaker state that the probability mass function associated with random variable X is defined as $f: R \rightarrow [0, 1]$ where $f(x) = P(X = x)$. See Grimmett and Stirzaker (2001), p. 46.
88. Russell (1918), p. 199.
89. Hamilton (1994), p. 45.
90. Lindsay and Margenau (1963), p. 522.
91. Massumi (2010), p. 1.
92. Massumi (2015), p. 5.

93. Massumi (2015), p. 191.
94. Massumi (2015), pp. 42, 107, 137.
95. Massumi (2015), p. 194.
96. Massumi (2015), p. 203.
97. Massumi (2015), p. 13.
98. Avanessian and Malik (n.d.).
99. Massumi (2015), p. 13.
100. An iterated modality traps a modal operator within the scope of another, so that two or more such modal operators occur in the relevant formula.
101. Massumi (2015), p. 3.
102. Schmitt and Stevenson (2004).
103. Massumi (2015), p. 191.
104. Massumi (2015), p. 191.
105. Massumi (2015), p. 190.
106. Trzęsicki (2015), p. 140.
107. Trzęsicki (2015), p. 141.
108. Trzęsicki (2015), p. 141.
109. Trzęsicki (2015), p. 143.
110. It should be noted that addition of this branching structure is not uncontroversial in discussions of temporal logic. See e.g. Goranko and Rumberg (2020).
111. Trzęsicki (2015), p. 143. More fully, Trzęsicki claims that in branching temporal logic he sets out, the statement that a proposition will be true is determined if, and only if, for all branches, it will be true at some future moment. 'it will be φ ' is determined iff on each branch at some moment in the future it will be φ.
112. Trzęsicki (2015), pp. 143–144.
113. Trzęsicki (2015), p. 144.
114. Massumi uses terminology similar to the images of branching to describe the modality of future moments, referring to future moments cutting into the present, co-presences of moments with other moments, and spatialized modes of time, conceived as convergent and divergent series of paths. See Massumi (2015), p. 117.
115. Trzęsicki (2015), p. 144.
116. Goranko and Rumberg (2020).
117. Formally, $T = \{1, 2\}$, $I = \{a, b\}$, $1 < 2$, $p \in V_a(1)$, $p \in \mathrm{V}_{\mathrm{b}}(1)$, $(2, \mathrm{a}) \lhd (2, b)$, but it is false that $(1, a) \lhd (1, b)$. Trzęsicki (2015), p. 157.

118. More fully,we have that $\langle T, <, \lhd, V_i \rangle, t \vDash \Box \varphi$ iff for any j : if $(t, i) \lhd (t, j)$, then $\langle T, <, \lhd, V_j \rangle, t \vDash \varphi$. Take $i = a$, $t = 1$. Since $(1, a) \lhd (1, b)$ fails, $\langle T, <, \lhd, V_a \rangle, 1 \vDash \Box \varphi$. However, the definition of P requires that $\langle T, <, \lhd, V_i \rangle, t \vDash P\varphi$ iff there is $t_1 \in T$, $t_1 < t$ such that: for any $j \in I$: if $(t, j) \lhd (t, i)$, then $\langle T, <, \lhd, V_j \rangle, t_1 \vDash \varphi$. For $i = a$, $t = 1$, there is no such t_1. $<t$, so whilst $\langle T, <, \lhd, V_i \rangle, t \vDash \Box \varphi$ holds $\langle T, <, \lhd, V_i \rangle, t \vDash GP\varphi$ does not. Thus $\Box\varphi \rightarrow GP\varphi$ fails.
119. Kant (1933), A426/B454.
120. Everettian quantum mechanics offers a deterministic account of the world's evolution, which takes place in accordance with the Schrödinger equation, as a superposition of basis states, each of which is taken to correspond to an independent world. The universe's evolution thus takes on a branching structure. See e.g. Wallace (2012).
121. Massumi (2015), p. 191.
122. Massumi (2015), p. 3.
123. Schmitt and Stevenson (2004).
124. Massumi (2015), p. 191.
125. Massumi (2015), p. 190.
126. Trzęsicki (2015), p. 141.
127. Massumi (2015), p. 194.
128. Massumi (2015), p. 137.
129. Massumi (2015), p. 117.
130. Deleuze (2004), pp. 89–93.
131. Coinmarketcap.com.
132. https://ledgerjournal.org/ojs/ledger.
133. Nakomoto (2008b).
134. Land (2018), p. 1356.
135. In particular, the claim is that protocols employed by blockchain-instantiating cryptocurrencies provide an "absolute" basis for assigning temporal coordinates to events.
136. Lanksy (2018), p. 19 defines incorporation of distributed consensus as one of six marks of cryptocurrencies.
137. Eyal and Sirer (2013), p. 1.
138. Narayanan et al. (2015), p. 14.
139. Narayanan et al. (2015), pp. 25–26.
140. Back (2002).
141. Narayanan et al. (2015), p. 39.
142. Narayanan et al. (2015). p. 39.

143. Nakomoto (2008a), p. 2.
144. This said, an actor able to command a significant amount of the computing power on the network could succeed in defrauding others, by ignoring newly mined blocks which contains transactions they wish to reverse (such as them remitting currency to another party) and begin mining from the latest block which does not include such a transaction. The probability of successfully mining enough blocks without being beaten to the punch by another miner scales with the available computing power. Some species of these attacks, sometimes deemed block-withholding attacks involve the party successfully mining a block and deliberately failing to broadcast it, with the hope of discovering several consecutive blocks, independent of the rest of the chain. If successful in mining these blocks quicker than the communicative nodes, the saboteur could then declare them all at once, departing from an earlier block in the chain, overtaking and thus superceding the chain with which other miners are preoccupied, "orphaning" any blocks mined by them in the interim. Narayanan et al. (2015), p. 73.
145. Nakomoto (2008a), p. 3.
146. Nakomoto (2008b).
147. Nakomoto (2008a), p. 1.
148. Nakomoto (2008a).
149. Indeed, going back further still, problems of synchronisation were fundamental long before the advent of modern computing and the development of systems theory. Peter Galison's text *Einstein's Clocks, Poincaré's Maps: Empires of Time* presents a comparative history of the two physicists' careers, and attempts to locate their innovations in electrodynamics in the socio-political context of nineteenth and twentieth Century Europe, together with the practical rather than theoretical engineering challenges which beset both of them in their extra-academic lives. See Galison (2004).
150. Kangasharju (2013), p. 8.
151. See e.g. Turner (2007), Lamport (1998, 2001, 2005).
152. Lamport (2001), p. 2.
153. Turner (2007), pp. 3–4.
154. Lamport (2005), p. 1.
155. Turner (2007), p. 16.

156. More fully, Narayanan et al. identify "four checks" performed by nodes as necessary but not sufficient conditions for acceptance of a broadcast transaction. See Narayanan et al. (2015), p. 72.
157. Wood (2018), p. 2.
158. Bonneau et al. (2015), p. 10.
159. Lamport indulges a broader metaphysical hypothesis, going so far as to claim that time is derivative from event-ordering. Lamport (1978), p. 558.
160. Narayanan et al. (2015), p. 72.
161. Of course, this assumes that the credit line available to the participants is a parsimonious one.
162. Strictly speaking, it is mistaken to equate events, transactions and the formation of blocks. There cannot be a one-for-one correspondence of the latter to the former, since blocks embed a large number of transactions, taking a significant amount of time to mine. At the beginning of 2020, this time averaged out at around ten minutes: see e.g. Narayanan et al. (2015), p. 45, or https://data.bitcoinity.org/bitcoin/block_time/.
163. For a mathematical illustration of this see Nakomoto (2008a), pp. 6–8.
164. Narayanan et al. (2015), p. 144.
165. Nakamoto (2008a), p. 5.
166. Brassier et al., (2017).
167. Poincaré (1913), p. 234.
168. Brown (2005), p. 19.
169. This principle has been referred to as the correspondence principle and is occasionally (controversially) attributed to the work of Niels Bohr. See e.g. Bokulich and Bokulich (2020).
170. Albert (1992), pp. 43–44.
171. Hesse (1962), p. 152.
172. Russell's concern, for example, is with whether "there is a functional relation of the form" specified. Russell (1918), p. 199.
173. Peirce (1891).
174. Peirce (1982–1993), p. 293.
175. Smolin (2013), p. 123.
176. Smolin (2013), p. 97.
177. Smolin (1997), p. 106.

Bibliography

Abbott, A. (2001). *Time Matters: On Theory and Method*. University of Chicago Press.

Abbott, A. (2016). *Processual Sociology*. University of Chicago Press.

Albert, D. (1992). *Quantum Mechanics and Experience*. Harvard University Press.

Anderson, D., & Zalta, E. (2004). Frege, Boolos, and Logical Objects. *Journal of Philosophical Logic, 33*(1), 1–26.

Avanessian, A., & Malik, S. (n.d.). The Time-Complex. *DIS Magazine*. Retrieved September 26, 2019, from http://dismagazine.com/discussion/81924/the-time-complex-postcontemporary/

Ayache, E. (2010). *The Blank Swan: The End of Probability*. John Wiley & Sons.

Ayache, E. (2015). *The Medium of Contingency: An Inverse View of the Market*. Palgrave Macmillan.

Ayache, E. (2016). Response to Johnson: A Random Sample Versus the Radical Event. *Finance and Society, 2*(2), 205–216.

Back, A. (2002). *Hashcash—A Denial of Service Counter-Measure*.http://www.hashcash.org/papers/hashcash.pdf

Barbour, J. (1989). *Absolute or Relative Motion? Vol. 1: The Discovery of Dynamics*. Cambridge University Press.

Berkovitz, J., Frigg, R., & Kronz, F. (2006). The Ergodic Hierarchy, Randomness and Hamiltonian Chaos. *Studies in History and Philosophy of Modern Physics, 37*, 661–691.

Bokulich, A., & Bokulich, P. (2020). Bohr's Correspondence Principle. In E. Zalta (Ed.), *The Stanford Encyclopedia of Philosophy*. https://plato.stanford.edu/archives/fall2020/entries/bohr-correspondence

Bonneau, J., et al. (2015). *SoK: Research Perspectives and Challenges for Bitcoin Cryptocurrencies*. IEEE Symposium on Security and Privacy. http://www.jbonneau.com/doc/BMCNKF15-IEEESP-bitcoin.pdf

Brassier, R., et al. (2017). Pricing Time: Outline and Discussion on Suhail Malik's "The Ontology of Finance". *Identities: Journal of Politics, Gender and Culture, 14*(1), 84–128.

Brock, W., & De Lima, P. (1996). Nonlinear Time Series, Complexity Theory, and Finance. In G. Maddala & C. Rao (Eds.), *Handbook of Statistics Volume 14: Statistical Methods in Finance*. Elsevier Science.

Brown, H. (2005). *Physical Relativity: Space-Time Structure from a Dynamical Perspective*. Oxford University Press.

Campbell, J., Lo, A., & MacKinlay, C. (1997). *The Econometrics of Financial Markets*. Princeton University Press.
Chatfield, C., & Xing, H. (2019). *The Analysis of Time Series*. Taylor & Francis.
Deleuze, G. (2004). *Difference and Repetition* (P. Patton, Trans.). Continuum.
Esposito, E. (2011). *The Future of Futures: The Time of Money in Financing and Society* (E. Esposito, Trans.). Edward Elgar.
Esposito, E. (2018). Predicted Uncertainty: Volatility Calculus and the Indeterminacy of the Future. In J. Beckert & R. Bronk (Eds.), *Uncertain Futures: Imaginaries, Narratives, and Calculation in the Economy* (pp. 219–235). Oxford University Press.
Eyal, I., & Sirer, E. (2013). *Majority Is Not Enough: Bitcoin Mining Is Vulnerable*. https://arxiv.org/abs/1311.0243
Fama, E. (1970). Efficient Capital Markets: A Review of Theory and Empirical Work. *The Journal of Finance, 25*(2), 383–417.
Galison, P. (2004). *Einstein's Clocks, Poincare's Maps: Empires of Time*. Hodder and Stoughton.
Goranko, V., & Rumberg, A. (2020). Temporal Logic. In E. Zalta (Ed.), *The Stanford Encyclopedia of Philosophy*. Retrieved August 7, 2020, from https://plato.stanford.edu/archives/sum2020/entries/logic-temporal/
Grimmett, G., & Stirzaker, D. (2001). *Probability and Random Processes*. Oxford University Press.
Hamilton, J. (1994). *Time Series Analysis*. Princeton University Press.
Hesse, M. (1962). *Forces and Fields: The Concept of Action at a Distance in the History of Physics*. Dover Publications.
Husbands, C. (2020a). Esposito's Temporality of Finance: Endogeneity and Revisability in Derivative Transactions. *Finance and Society, 6*(2), 114–129.
Hull, J. (2002). *Options, Futures and Other Derivatives* (5th ed.). Prentice Hall.
Johnson, T. (2016). The Necessity of Multi-disciplinary Scholarship for Finance: On Ayache and Roffe. *Finance and Society, 2*(2), 189–204.
Kangasharju, J. (2013). *Distributed Systems: Synchronisation*. University of Helsinki. Retrieved August 23, 2020, from https://www.cs.helsinki.fi/webfm_send/1232
Kant, I. (1933). *Immanuel Kant's Critique of Pure Reason* (N. Smith, Trans.). Macmillan.
Keynes, J. (1936). *The General Theory of Employment, Interest, and Money*. Macmillan. Retrieved June 8, 2020, from https://www.files.ethz.ch/isn/125515/1366_KeynesTheoryofEmployment.pdf

Lamport, L. (1978). Time, Clocks and the Ordering of Events in a Distributed System. *Communications of the ACM, 21*(7), 558–565.

Lamport, L. (1998). The Part-Time Parliament. *ACM Transactions on Computer Systems, 16*(2), 133–169.

Lamport, L. (2001). Paxos Made Simple. Retrieved August 24, 2020, from http://lamport.azurewebsites.net/pubs/paxos-simple.pdf

Lamport, L. (2005). Generalized Consensus and Paxos. Retrieved August 24, 2020, from https://www.microsoft.com/en-us/research/wp-content/uploads/2016/02/tr-2005-33.pdf

Land, N. (2018). *Crypto-Current, An Introduction to Bitcoin and Philosophy*. Retrieved August 22, 2020, from http://sumrevija.si/wp-content/uploads/2018/11/PDF.pdf

Lansky, J. (2018). Possible State Approaches to Cryptocurrencies. *Journal of Systems Integration, 9*(1), 19–31.

Lindsay, R., & Margenau, H. (1963). *Foundations of Physics*. Dover Publications.

Luhmann, N. (1981). Temporalstrukturen des Handlungssystems. In *Soziologische Aufklärung; 3. Soziales System, Gesellschaft, Organisation*. Westdeutsche Verlag.

MacKenzie, D. (2006). *An Engine, Not a Camera: How Financial Models Shape Markets*. MIT Press.

Malik, S. (2014). The Ontology of Finance. In *Collapse VIII* (pp. 629–812). Urbanomic.

Margenau, H. (1977). *The Nature of Physical Reality: A Philosophy of Modern Physics*. Ox Bow Press.

Massumi, B. (2010). Perception Attack: Brief on War Time. *Theory and Event, 13*, 3.

Massumi, B. (2015). *Ontopower: War, Powers, and the State of Perception*. Duke University Press.

Nakomoto, S. (2008a). *Bitcoin: A Peer-to-Peer Electronic Cash System.*https://bitcoin.org/bitcoin.pdf

Nakomoto, S. (2008b). *Re: Bitcoin P2P E-Cash Paper*. https://satoshi.nakamotoinstitute.org/emails/cryptography/11/

Narayanan, A., et al. (2015). *Bitcoin and Cryptocurrency Technologies (Draft)*. Princeton University. Retrieved January 2018, from http://www.theblockchain.com/docs/Princeton%20Bitcoin%20and%20Cryptocurrency%20Technologies%20Course.pdf

Newton, I. (1999). *The Principia: Mathematical Principles of Natural Philosophy* (I. Cohen & A. Whiteman, Trans.). University of California Press.

Nietzsche, F. (1967). *The Will to Power* (W. Kaufmann & R. Hollingdale, Trans.). Random House.
Peirce, C. (1891). The Architecture of Theories. *The Monist, 1*(2), 161–176.
Peirce, C. (1982–1993). *Writings of Charles S. Peirce: A Chronological Edition, Volumes 1–5*. Indiana University.
Poincaré, H. (1913). The Measure of Time. In *The Foundations of Science (The Value of Science)* (pp. 222–234). Science Press.
Robinson, J. (1981). *What Are the Questions? And Other Essays: Further Contributions to Modern Economics*. M. E. Sharpe.
Roffe, J. (2015). *Abstract Market Theory*. Palgrave Macmillan.
Russell, B. (1918). *Mysticism and Logic and Other Essays*. Longmans, Green and Co.
Schmitt, E., & Stevenson, R. (2004). Admitting Intelligence Flaws, Bush Stands by Need for War. *New York Times*, July 10, A9.
Smolin, L. (1997). *The Life of the Cosmos*. Oxford University Press.
Smolin, L. (2013). *Time Reborn*. Penguin.
Soros, G. (1987). *The Alchemy of Finance: Reading the Mind of the Market*. Wiley.
Trzęsicki, K. (2015). Indeterministic Temporal Logic. *Studies In Logic, Grammar and Rhetoric, 4*(1), 139–162.
Turner, B. (2007). The Paxos Family of Consensus Protocols. Retrieved August 24, 2020, from http://www.fractalscape.org/files/paxos-family.pdf
UBS Chief Investment Office. (2020). *UBS Global Real Estate Bubble Index 2020*. UBS Switzerland AG.
Wallace, D. (2012). *The Emergent Multiverse: Quantum Theory According to the Everett Interpretation*. Oxford University Press.
Wood, G. (2018). *Ethereum: A Secure Decentralised Generalised Transaction Ledger*.https://ethereum.github.io/yellowpaper/paper.pdf

4

Temporality of Determinacy II: Philosophy of Physical Sciences

Introduction

Its sporadic appearance in millennia of philosophical discussions might lead one to take the temporality of laws as a problem removed from the frontiers of physical science. Surfacing in Ancient debates over logic and predestination and resurfacing in Twentieth Century metaphysics, it has been the subject of exposition for its own sake, rather than appearing in philosophy only as a translation of the conclusions forced through by scientific research and outcomes of experiment. Clearly, the reflections of Cicero, Chrysippus and Plutarch, as well as the interventions of the scholastic theologians many Centuries later, befit the former classification. But such a conclusion applies also to more recent contributors to the question. C. S. Peirce, the great American logician and one of the field of semiotics' fathers, reflected extensively on the temporal status of laws and embedded his discussions in his sceptical attacks on determinism. According to T. Alborn's account of Peirce, the latter embraced an evolutionism with respect to physical laws and, prompted by his studies of Darwinian biology, extended this analysis to *all* existential categories, not only nature, including logic itself.[1] In so doing Peirce constituted what he terms a Cosmogonic Philosophy in which nature evolves in and with the natural world. Shortly after Peirce's contributions, and with his early

C. Husbands, *The Temporality of Determinacy*,
https://doi.org/10.1007/978-3-030-86530-6_4

career dedicated to the foundations of mathematics and logic, the mathematician-philosopher A. N. Whitehead turns to the metaphysics of determinacy in texts such as *Process and Reality* and *Science and the Modern World*. Whitehead makes a number of claims which align him closely to the evolutionary vision of Peirce: he speaks of physical laws being only approximately obeyed, subject to transitions over time, and gradually coming into effect.[2] He refers to the laws of nature evolving into being as patterns and regularities[3] rather than as timeless or atemporal forms. Whitehead's treatise sets out a broad process philosophy rather than an empirical one which takes off directly from the fruits of scientific research. Whitehead and Peirce are far from isolated cases, and similar perspectives have emerged from distinguished thinkers in the philosophy of science. Paul Feyerabend, for example, in his "Realism and the Historicity of Knowledge," dismisses time-invariant and time-independent laws as theological relics in favour of his now well known methodological pluralism.[4]

The results of the previous chapter might prompt similar conclusions. In the non-physical sciences, the role of systems of participants, reciprocal expectations, and actors and networks in the temporality of determinacy might lead one to deem this a mere *phenomenological* oddity: in these fields, the conception of determinacy as a process—not to mention the other features of theories analysed in the previous chapter—might be thought to reflect epistemic limitations rather than anything more fundamental. As mentioned in its conclusion, this suspicion manifests itself in the faith some commentators invest in the potential for the future to deliver knowledge of fundamental or underlying dynamics, and in the derogation of the rules and laws of the subject disciplines as provisional and approximate. According to this suspicion, advances in these disciplines will one day supercede this primitive conceptual arsenal, replacing it with a more granular theory free of such strange characteristics. In principle, one will one day be able to conjure a successor to Laplace's demon fit for modernity, and identify all fundamental variables determining the movements of participants in social systems, from the constituents of their C-fibres to the particles comprising adverse weather systems, measuring them exactly, and grasping their eternally determinate evolution over time. To the extent that such a theory is unavailable at present, the provenance of these elliptical temporal dependencies are considered to be nothing more than the unanalysed compound entities which are considered in these non-physical fields, such as utility maximizing agents, with their expectations and interactions, or perhaps the fact that such compounds are open systems which cannot be isolable in the way their constituents are in the ideal laboratory. They reflect nothing of the system's components. Such would be the way the philosophy of atemporality would categorize theories with time-dependencies of this nature.

As mentioned in the conclusion to the previous chapter, this situation is not in fact an eccentricity of the non-physical sciences. No uncontroversial notion of scale permits such a distinction in the physical sciences either, nor is there consensus on the prospects for an ultimate theory, couched in irreducible dynamics, to provide a clear example of what the terms "fundamental" or "underlying" mean. But even if one were to discard this reply, there is a more basic reason why discussions of the nexus of temporality and determinacy and the temporal status of laws are not confined to metaphysics, the philosophy of science, or the study of social systems, where functional relations are rarely considered germane to the debate. In fact, a number of scholars working in the physical sciences, whether as researchers or theoreticians, have contributed to the question. Recently, these contributions have appeared most frequently under the avatar of the *variation* or the *evolution* of the laws which confer determinacy. This variation or evolution of laws is one of the clearest ways in which temporal determinacy appears in the physical sciences, and will be an important topic featured in this chapter.

One example arises in the work of a particularly distinguished Twentieth Century commentator on the foundations of physics, John Wheeler. Wheeler was responsible for some of the most striking as well as philosophically fertile theoretical advances, encompassing everything from relativity to quantum mechanics—from the Wheeler-Feynman electromagnetic absorber theory to the quantum delayed-choice Gedankenexperiment (often adapted into the delayed-choice quantum eraser), with what he describes as its inversion of temporal order.[5] His more expansive *Frontiers of Time* as well as essays in *Quantum Theory of Measurement* offer reflections on the temporal status of laws or, in his elusive styling, law without law. In these texts, the possibility of time-dependent laws is explored through the lenses of both quantum mechanics and general relativistic cosmology. In these he asks explicitly whether physical laws are eternal, immutable and unchanging or whether, instead, they are like species, mutable, evolving and the product of fortuity and randomness.[6] Wheeler insists that accounting for the emergence of laws should be a task as essential to science as explaining anything else, from the constitution of substances to space-time.[7] The prospect of eternal laws, constrained from the universe's inception (whether or not engineered by an intelligence) he dismisses aggressively, insisting on their derivative and secondary, as opposed to fundamental and primary character.[8]

Wheeler's argument for the mutability of laws traverses at least three main paths in these texts. First of all, his essay on laws without laws considers the Wheeler delayed choice experiment (to be discussed later in this chapter in more detail), scaled up to cosmological rather than laboratory scale, and its implications for the distinction between modes of time. Wheeler speaks of measurements performed having effects on or, indeed, effecting altogether states of affairs occurring earlier in time, resonating with the distinction established in previous chapters between the order of occurrence and order of determination attributed to events, an ostensible implication of a number of quantum mechanical measurement processes which is extremized by the staging of the delayed-choice or quantum eraser scenario. Wheeler speaks of the usurpation of the normal order of time, present experiments and observations having consequences for microscopic particles eons prior to the emergence of life, and of what he describes as a creative act reaching back into present time from literally billions of years prior.[9] In this discussion, then, Wheeler anthropomorphizes present measurements' determination of past events both as the past reaching into the present and the present into the past, attesting to the conceptual difficulties organising relationships between different modes of time directionally. Present measurements determine past events, reaching one way; equally though, measurements construct or reconstruct the evolution from the past to the present act of measurement. Wheeler at times suggests a species of retrocausality linking past and present. Either way, Wheeler suggests that just as previous events depend on later events, so do the laws on which these events depend, and in respect of which they are determined. The laws themselves are subject to time in the same sense as are events.

Second of all, Wheeler draws a surprising link between two less obvious characteristics and the evolutionary status of laws: (i) the fact that laws are approximate, and (ii) the fact that laws are statistical. This he suggests stems from the observer participancy[10] he finds in contemporary physics, particularly in orthodox[11] formulations of quantum mechanics which and whose metaphysical implications encompass the mutability of laws.[12] In *Frontiers of Time*, Wheeler is even more explicit: all legitimate laws must be statistical, and being statistical is a sufficient condition for being changeable.[13] With respect to the Permanence Thesis, this is a gauntlet thrown down. This foundation, which consists of uncountable numbers of measurements and revisions to expectations which reverberate over eons, produces a patchwork quilt of laws which shift with the scale and sophistication of the experiments which test them.

Thirdly and finally, Wheeler makes a number of explicit statements as to the evolution of laws, motivated not by quantum experiments extrapolated to the cosmological scale, but by cosmology proper: the physics of the early universe came to be with the big bang and the birth of spacetime itself, hence his derogation of the rigidity of laws of nature, fashioned either by intelligent design or by whatever other mechanism might see them hold from what he describes as eternity to eternity.[14] Instead, he claims the laws must have come to be, at the very beginning only ventriloquizing quantum perturbations, before the relative uniformity and consistency of nature emerged like foam at a statistical scale.

Wheeler is not alone in proclaiming the evolutionary resonances of laws, nor in his ambition to link this trait to aspects of modern cosmology. Other physicists have followed similar steps, going as far as to posit natural selection mechanisms applying at the cosmological or supra-cosmological scale to distinct universes with distinct laws, and providing quantitative detail and logical steps where Wheeler at times (e.g. in *Frontiers of Time*) leaves the connection between the provenance of the universe and the impermanence of laws unelaborated. The next section considers some of the most prominent ways in which this proposal has been fleshed out in more detail in Twentieth Century contributions to cosmology. It also considers the implications of these evolutionary nomological models for the atemporality conjecture. It then turns to other areas of the physical sciences in which recent developments refute aspects of this conjecture, including statistical mechanics and quantum theory.

The Evolution of Laws: Cosmology

As can be seen from this introduction, cosmology is among the fields of physics which Wheeler associated closely with time-dependency. Others, such as Whitehead and Peirce, share this vision, with their more general nomological evolutionism. Another important recent contributor is Lee Smolin. Smolin's primary impetus in published tracts such as *The Life of the Cosmos*, *Time Reborn* and *Einstein's Unfinished Revolution*, as well as in his voluminous academic papers on the subject, presents laws as structures which *emerge* from the regularities fostered by the transition from the universe's birth to epochs of stability. In his 1992 paper "Did the Universe Evolve?" Smolin provides a more comprehensive theoretical framework for the emergence of these laws, which considers a multitude of universes, each of whose laws are parameterised by different constants, distinguished by random fluctuations. The unusual quantum of and relativities between these parameters which are functionally determinative of the laws in which they appear—from the gravitational constant G to the ratio of the masses of neutrons and protons—is as inexplicable as is the origination of the universe. The fortuitously fecund conditions for the universe's expansion, the synthesis of heavier molecules, the existence of habitable planets, and the eventual upspring of intelligent life, depend sensitively on these parameters falling within narrow a range—accounting for which conspicuous fluke requires, for Smolin, the postulation of a natural selection mechanism which holds over a population of universes. Over this population, this mechanism effectively selects for universes which generate increased numbers of black holes, since the greater part of the space of possible universes, distinguished by the parameters they contain, and the generation to which they belong, that is, how many previously collapsed universes they follow after, is occupied by universes with parameters conducive to the production of black holes and with greater progeny. Smolin refers to this as a form of natural selection at the cosmological scale; parameters which produce larger numbers of black holes are those which are selected for over time.[15]

Such a hypothesis is, whilst avowedly speculative, not unempirical according to Smolin. At least, he claims that it is falsifiable.[16] His approach, therefore, subjects not just the universe but the parameters that determine its fundamental physics[17] to evolution over time. Insofar as these parameters parameterize laws, and therefore represent dynamics, it furnishes a sense in which dynamical laws vary in time—and thus aspires to fulfil Peirce's ambition of resolving laws into a deeper evolutionary mechanism.[18] Of course, with its expanded focus on vast numbers of universes rather than one, this move requires drawing a distinction between the different time-scales inhabiting or inhabited by each. Just as in the context of the non-physical sciences a distinction was introduced between the time t featured in pricing models (the time in which, for instance, transactions are executed) and on the other hand the time τ in which expectations about the attributes of these transactions (such as price levels and their dispersion) are formed and revised, here the prospect of different universes instantiating different dynamics necessitates a distinction between the time of a given universe, whose symbol appears in the dynamical laws proper to it, and the larger time scale over which generations of universes are born and die. In this sense, Smolin's hypothesis thrusts out the frontiers of time back before the beginning of the universe, after its putative collapse into singularity, and outwards through every other episode of black hole formation in its lifespan.

Of course, there is a demonstrable and direct clash between this line of thought and the atemporality theses analysed throughout this work. Variation of the parameters of basic laws amounts to variation in these laws, with dramatically different trajectories emerging for the evolution of the universe if these parameters undergo even seemingly minor perturbations. To the extent that the time in respect of which the permanence of laws is judged stretches beyond a single universe's lifespan, the eponymous thesis fails for Smolin's model. Further, Smolin's focus on the import of changing parameters to determinacy is no idiosyncrasy; other commentators on the temporality of determinacy have asserted this link, including those who make constancy of parameters in functional laws necessary conditions for consistency or for causal relations to obtain. Take, for illustration, Lindsay and Margenau's *Foundations of Physics*: the consistency of nature requires that the equations describing nature do not depend on time in what the authors regard as an absolute manner, and the transformation $t \rightarrow t_0$ transforms solutions of these equations to solutions of these equations.[19] As a result, they claim, the parameters appearing in these laws, such as the exponent in Coulomb's law of attraction in electromagnetism, are "invariable in time" and remain eternally the same.

Yet such exponents are precisely the constants rendered inconstant by cosmological natural selection. Functions such as those which Smolin symbolizes $T(p)$ and $R(p)$, denoting respectively the expected lifetime of a universe and the expected number of final singularities in the universe, depend on the parameter set p. With variation in p from universe to universe, and therefore with cosmological or supra-cosmological time, such functional relations change. Similarly, as for the TVT approach to atemporality, under the mapping $t \rightarrow t + t_a$, subject to the condition that $t_a \sim T$ and thus that roughly an expected lifetime of a universe separates the times, p may vary, meaning solutions to dynamical equations under one parameter set, picking out a particular evolutionary trajectory, need not map to solutions of these equations once their fundamental constants are reshuffled by Smolin's hypothesised random fluctuations. The relationships between positions and velocities of charges interacting in accordance with the Coulomb rule could violate this law in another universe in which they were differently parameterised. As to the TVT, the implications of Smolin's approach are more subtle. His model offers few general remarks about occurrences of t in dynamical equations, nor any counterexamples to the claim that dynamical laws should be free of such occurrences. There is, though, an obvious sense in which these laws' very constituents might be describable or redescribable in a way which makes their implicit time-dependency explicit: Smolin might perhaps be inclined to suggest that parameters varying from universe to universe be indexed by some other variable such as τ, were it possible for us to surmount the anthropic limitations conveyed by being within one universe among others to not only distinguish between the different regimes of laws, but to associate them with different universes and different generations. Finally, in the case of the TVC, were this additional indexation introduced, whilst occurrences of $t_l \in \{t_i\}$ may not effect changes in the truth-value of the f in question with variation in t_l, this may be achieved by variations in τ_l instead. It would be fair to conclude Smolin's model is directionally in conflict with many of the core forms of the atemporality conjecture.

This aside, there is another interesting parallel between Smolin's work and that of other thinkers in the area of temporality and determinacy. Smolin asserts a logical connection between the evolution of laws and a causal closure principle which he alleges covertly underlies many branches of the physical sciences, unrecognized and unacknowledged by the greater part of the scholarship. In *Time Reborn* he refers to this as the cosmological fallacy, namely the temptation to take out of context a particular principle, having been successfully applied at a smaller scale to one of the subsystems of the world, and applying it then to the overall universe.[20] He diagnoses many of physics' unresolved conundrums and intra-theoretical tensions as symptoms of this fallacy, from quantum gravity to particle physics. Meanwhile, showing up another point of similarity between Smolin's work and the thinking of Lindsay and Margenau, *Foundations of Physics* draws various connections between closure and causality, or indeed time-dependency and the openness of the physical system being studied. In particular, the authors chalk up the presence of the time-variable in physical equations to incomplete analysis of the relevant system; counter-examples to the TVT cropping up simply means that the relationship has failed to take into account all the nontemporal variables at play, arising from extraneous forces. An analysis with a wide enough scope would (they contend) resolve superficial time-dependency into static underlying forces.[21] Time-dependency implies openness; closure requires atemporality –logical arrows common to both these authors and Smolin, but with only the latter taking them as grounds to embrace the conjunction of time-dependency and openness.

Smolin's model requires the unparsimonious postulation of innumerable parallel universes, differentiated by random fluctuations in their fundamental physical parameters, in order to give sense to the notion of the evolution of laws and the time-dependency of supposed constants. This could be conceived as a weakness. However, less far afield, the Twentieth Century has seen research dedicated to the possibility that evidence from this universe, in the form of live messages from the deep cosmological past, offer evidence for time-dependency of these parameters. Examples analysed by modern astronomy range from the constant G which appears in Newtonian gravitation and, more recently, general relativity, to the ubiquitous dimensionless fine-structure constant α. In their paper "Time-Variation of Newton's Constant and the Age of Globular Clusters," Degl'Innocenti et al discuss the effects of a variable G on the burn rate of stars, celestial orbits and stellar evolution, declaring in understated fashion that Newton's constant G could vary over cosmological epochs, and thus amount to a dynamical degree of freedom.[22] John Barrow, in his paper "Time-Varying G" considers Newtonian models which incorporate a gravitational constant which weakens gravity with time,[23] whereas his collaboration with J. Magueijo and H. Sandvik "A Cosmological Tale of Two Varying Constants"[24] considers the relation between the relative rates of variation of G and of α. Webb et al., in their 1999 "Search for Time Variation of the Fine Structure Constant" use spectroscopy to gather evidence which they suggest is consistent with varying α. These investigations into the possible variability of G, of α, of the ratio of the electron and proton masses, of the Higgs vacuum expectation value and of many other parameters, have engendered a rich tapestry of academic debate about these and other forms of parameter variability.[25]

As early as the 1930s—long before the deployment of the precise techniques of contemporary astronomy—Paul Dirac's writings had already anticipated these principles, positing the evolution of parameters with the expansion of the universe. Dirac's starting point was what he took to be a conspicuous equivalence—of the powers of ratios of these parameters, whence he claimed to derive from first principles that some of the very large numbers involved vary between cosmological epochs.[26] The ratio γ of the electric force between an electron and a proton to the gravitational force between them, and the ratio of the age of the universe to the atomic light-crossing time $\frac{e^2}{m_e c^3}$, coincide up to a constant of order unity. In view of this elliptical coincidence, Dirac postulated an undiscovered physical law which would account for the equality of the two ratios, and tied the order of magnitude involved back to the age of the universe. One of the decisive consequences of this hypothesis for the question of temporal determinacy is the implication that, in order for the equivalence between these ratios to be preserved, as the universe ages, γ must increase proportionally. As time elapses, the ratios scale in tandem, thus all very large dimensionless numbers in nature, are powers of epochal time, with coefficients of the order unity.[27] In turn, this is the origin of the thesis that the gravitational constant must vary with the passage of great intervals of time,[28] decreasing the denominator of γ and confirming Dirac's model. Remarkably, Dirac considers the possibility of introducing an additional time variable τ in terms of which t itself can be parameterised, related as $\tau = \log(t)$. Transformation to this new variable accommodates the normal forms of the laws of mechanics in spite of fluctuations in the gravitational constant G with t.[29] Of course, the parallels here with the content of previous chapters—albeit with different reasons and in different contexts—are striking: the pricing relations subject to revision in Esposito's theory of financial markets provide just one apposite example. As will be seen later, other physicists, including quantum researchers such as John Cramer, have posited similar notions of pseudo-time in which processes which are not circumscribed by the dynamics of the theory impart determinacy to physical systems (in Cramer's case, through the transaction of forwards- and backwards-propagating electromagnetic radiation).[30]

Dirac's innovative combination of dimensional and scale analysis with cosmological concepts has led him to be credited by several contributors to the debate over varying constants as being among the leading progenitors of this area of research. For one, Kafatos, Roy and Roy mention Dirac alongside Edward Milne and Arthur Eddington as being the first to consider the gravitational constant as a time-varying quantity.[31] Dzuba, Flambaum and Webb hold Dirac responsible for the first reflections on how fundamental physical constants might have varied with the expansion of the Universe.[32] Barrow and Tipler, in their compendium *The Anthropic Cosmological Principle*, trace numerological reasoning (comparable to Dirac's) back through millennia to Archimedes, but acknowledge the originality of Dirac's decision to make the gravitational constant the host for the time-variation adduced to explain conspicuously similar relativities.[33]

How can these insights be brought back around to the nexus of temporality and determinacy? Despite that the variability of physical constants in the case of Dirac's thinking arises as an artefact of the evolution of our own universe, rather than the transition between different universes, the consequences are overall similar to those of Smolin's theory: time-dependent physical constants signify time-dependent laws. Thus, if these perspectives on the variability of fundamental constants were embraced, little further work would be required to invalidate the framing of physical laws found in the atemporality conjecture.

Revisit the various forms of the conjecture, in order to see this more clearly. The original version of the TVT was the following:

> For no law f does the time-variable t occur in f where $ag'' + bg' + cg + d = f$ and $g' = \frac{\partial g}{\partial t}$, $g'' = \frac{\partial^2 g}{\partial t^2}$. (Time-Variable Thesis)

By way of just one straightforward rejoinder, the time-dependency of the fine-structure constant, which is related to the Coulomb constant by the equation $k_e = \frac{hc\alpha}{2\pi e^2}$, undermines the time-independence of the Coulomb relation. The relevant differential equation befitting the TVT, $F = \frac{dp}{dt} = k_e \frac{q_1 q_2}{r^2}$, which fulfils the above schema for the combination $d = 0$, $c = 0$, $d = 0$ and for the case in which f is Coulomb's formula, contains a time-dependent k_e. Simply changing the way in which f were expressed, such as making its time-dependency explicit, would lead to a discernible violation of the TVT.

However, the consequences for the principle of translation invariance are less clear. In places, Dirac gives the impression that his manoeuvring is designed precisely to *maintain* (rather than violate) relations between fundamental physical constants over time. In such a case, transformations through time would not invalidate sets of solutions of dynamical laws. There is clear evidence for this intention in his 1938 text.[34] From this perspective, the reason for introducing a temporal dependency into physical constants is to ensure conspicuously similar ratios move in step, so as *not* to disturb the equations in which they appear. This would suggest that if the gravitational constant decreases with time as Dirac suggests[35] then a translation between epochs $t_{epoch\,A} \rightarrow t_{epoch\,B}$ will map solutions of laws to solutions—not to mention the other variables, besides the gravitational constant, which might fall within the purview of this mode of analysis (such as the cosmological constant or light speed). However, whilst maximum fidelity to the established functional relations between G, e, c etc. might well have been a desirable objective, Dirac's 1938 account is, in this sense, far from complete: there is scant demonstration that introduction of time-variation to these parameters does not lead to unintended distortions in the various physical theories which combine and permute them in various physically important laws, particularly in the quantum domain. Dirac's approach is in this respect a sketch rather than a theory. As Barrow and Tipler note in their analysis, it is quite an assumption that no other counterveiling changes need be made to ensure the coherence of these laws. Rates of nuclear fusion, solar luminosity and the radius of the Earth's orbit about the Sun, have all been claimed as physical processes sensitive to whether and how one takes varying G seriously—and all of them can be observed.[36]

Finally, how does the notion of permanence, taken by Lindsay and Margenau to be tightly bound up with the vaunted atemporality of laws, square with time-variable physical constants? The authors' statements open up diverging possibilities. On the one hand, time-varying constants would appear to violate the intuition that permanence requires the various components of dynamical laws to remain in fixed relationships, for constants to be constant with time, and for fundamental physical parameters to be at a remove from the theatre in which bodies interact—to purloin Donald Mackenzie's framing, they would be cameras rather than engines. On the other hand, if Dirac is to be believed, important ratios which relate fundamental forces and parameters to one another can be preserved in spite of this time-dependency. However constants may vary, the equations they comprise still exhibit the same structure, and still relate the same components, meaning there is a sense in which the impermanence of the quantum of the constants in the law need not imply the impermanence of the law itself.

In this way, these commentators, from Wheeler to Smolin to Dirac and his successors, assert a connection between time-dependency and questions of openness, closure and scale, designating cosmology a privileged source of insight into the evolution of laws. Whether this evolution occurs under the guise of the variation of physical parameters within this universe (Dirac et al.) or in the transition from one universe to another, the results are incompatible with many aspects of the atemporality conjecture. In the former case, this incompatibility relates primarily to the time-variable thesis; in the latter case, it stems primarily from permanence and translation invariance. Cosmology therefore becomes a conceptual furnace for the smelting of physical laws whose initially volatile and later imperceptible evolution provides a physical grounding for the temporality of determinacy, and the failure of metaphysical hypotheses as to their atemporality.

Temporality in Statistical Mechanics

Statistical mechanics has long served as a wellspring of metaphysical intrigue, and has engaged a diversity of philosophers as well as scientific practitioners. The gamut of problems which exist in the academic literature on the topic provide something of a microcosm of issues distinctive of the philosophy of science generally—problems relating to the reduction of macroscopic to microscopic theories, to the relation of dynamics to kinematics, to the consistency of properties such as chaos and stochasticity with determinism, and to the thermodynamic arrow of time.[37] Developed against the backdrop of thermal physics, with its study of gases, flows and cycles of heat which came to be deployed in the industrial revolution, statistical mechanics introduced probabilistic concepts into the discipline's foundations, predicting the evolution of variables in terms of distributions of their microscopic counterparts. Rather than attempting to compile knowledge of a complex multivariate system with copious degrees of freedom, such as a gas, from knowledge of the states of its constituents—a process not only intensive but subject to notorious objections—much more expedient probabilistic rules were developed to predict its overall behaviour, chiefly by Boltzmann and Gibbs, whose distinct approaches remain the canonical paths leading into the discipline. Accordingly, the theory's predictive or explanatory power derives not so much from classical dynamical laws applied to the smallest microscopic elements of a system, but from the constraints on the system implied by these probabilistic assumptions.

With only this brief staging, some of the philosophical problems are already visible. Can probabilistic explanations of complex systems' behaviour or empirical relations between variables such as temperature, pressure and volume ever boast the same explanatory power as a ground-up kinetic analysis of systems, which explains these variables in terms of changes in the positions and momenta of all of its particles?[38] Is there a sense in which the latter properties are more real, or a more legitimate object of investigation than the former, and could a reduction between the two be performed? Does the use of probabilistic concepts in the frameworks of Boltzmann and Gibbs reflect our ignorance about these microscopic states, or rather stochastic behaviour which is irremovably embedded in thermodynamic systems? What would such a notion of stochasticity—an altogether elusive one as Earman notes[39]—mean for determinism?

These questions provide only a perfunctory synopsis of the problems at stake. What is more, the problem with the least obvious relationship to the above description of thermodynamics is nonetheless among its most fervently debated aspects: namely, the uncertain provenance of time-asymmetry—more formally, time-reversal invariance—within the discipline. The systems described by thermodynamics are directional. Gases disperse evenly through empty space; organisms age and decompose over time; hot bodies transmit heat to their surroundings, tending to equilibrium. Nobody observes these quotidian occurrences running in reverse: spontaneous, convergent or self-organized processes are the subject of horror and fantasy, not empirical science. Yet this fact sits uncomfortably with the time-*symmetric* foundations of thermodynamics. The principles governing these systems' components (the billions of individual molecules vibrating in a gas, for instance) are consistent with processes happening both ways, and it is on these principles that their macroscopic evolution is typically regarded as being dependent. All of the motions of thermal systems' constituents consistent with its observed evolution over time are no more physically permissible than the reverse motions, just as a billiard ball could travel leftwards as well as rightwards across a surface with equal ease. However relevant, however expedient phenomenological or macroscopic descriptions may be, the systems they describe are composite. The behaviour of the composite is a function of the behaviour of its constituents, and the behaviour of these constituents leaves open a class of behavioural possibilities which are never explored by the composite. As Albert declares in his exemplary survey of the philosophical foundations of thermal physics *Time and Chance*, were all systems to evolve according to classical, Newtonian laws, no asymmetries could realistically arise between future and past.[40] This time-reversal invariance contrasts unusually with the consistent display of directionality on the part of so many independently evolving systems: whereas the laws of kinetic theory, governing as they do the behaviour of the microscopic constituents of thermodynamic systems, are time-reversal invariant, phenomena display a striking asymmetry. *Prima facie*, this is peculiar.[41]

Disagreements persist as to precisely where in the postulates or laws of thermodynamics time-asymmetry is derived—from the Second Law,[42] to observation,[43] to the very notion of equilibrium deployed by the theory, to the appeal of Boltzmann, Schrödinger and more recently Stephen Hawking to the psychological constraints of the observer of the theory's processes.[44] The frequently posited past hypothesis attempts to point to specific cosmological features of the early universe (namely those of extremely low entropy) in an attempt to explain consistent increases over time.[45] Boltzmann's H-Theorem has been presented as an attempt to derive constraints on thermodynamic systems from the behaviour of constituents, in the form of a constantly decreasing entropic function H.[46] What is less controversial is the claim that, somewhere in the historical and conceptual transition from the kinetic theory of gases to the probabilistic theory of statistical mechanics, some time-asymmetric ingredient must appear, however well disguised, in order to reconcile mechanics to experience. Otherwise, the everyday experience of processes occurring in one direction would remain unexplained by theory.

The problem of time-asymmetry is discussed very briefly here is relevant as a microcosm of wider philosophical problems in the field. It throws many of the previous questions into even more acute relief. This gives a sense of the rich linkages between thermodynamics and the philosophy of time, with its questions of symmetries and directions, and the way in which these are highlighted by the problem of temporal asymmetry in particular. It should come as no surprise, then, that the probabilistic architecture of statistical mechanics and the unidirectional processes it describes have ramifications for the temporality of determinacy, as the next section will elucidate with more detail.

Dynamical Systems Theory and the Ergodic Hierarchy

One of the most lucid frameworks relevant for statistical mechanics and its relation to time is provided by dynamical systems theory. The framework envisages systems as combinations of states transformed through time by a formal operator. Frigg, Berkovitz and Kronz characterise such a system as a triple $[X, \mu, T_t]$, with the following further definitions:

> $\{T_t \mid t$ is an instant of time$\}$ is a family of automorphisms, or transformations of X onto X
>
> T_t is such that $T_{t_1+t_2} = T_{t_1}\left(T_{t_2}\right)$ for all $x \in X$.[47]

This framing thus divides up a dynamical system into three components. X constitutes the phase space or state space of the system, the conjunction of all of the system's permissible vectors. In simpler terms, this space is the totality of possible states which can be occupied by the system, such as a distribution of some number of molecules in particular positions and with particular speeds. Each member of the space X corresponds to a state with given variables. For non-trivial systems, a system's evolution over time marks out a given phase space path or trajectory. This evolution over time is described by T_t, mapping states of the system onto states of the system; given a condition of the system at a given time, T_t determines its successors. In this sense, as far as determinacy is concerned, it resembles the more traditional notion of a law. This resemblance is underscored by the iterative property of the operator: from the property $T_{t_1+t_2} = T_{t_1}\left(T_{t_2}\right)$

it can be seen that, for discretely evolving systems, T_n can be decomposed into smaller factors, and ultimately to the repeated application of the initial operator. This implies the possibility of performing a technical reduction of any T_n to the *n*th iterate of T_1, T_1^n, grounding the idea that *one* law, rule or function is responsible for determining the entire evolution of the system. Further, the symbol μ represents the measure of the dynamical system. This measure assigns values to each region in the space and thus, more intuitively, amounts to the phase space equivalent of traditional measures of size. These elements constitute the basics of this framework for dynamical systems. It is in no way limited to statistical mechanics, but is instead general, and can be used to characterise classically deterministic systems as well as those which exhibit stochastic effects.[48]

Time is etched recognizably into the formalism of this framework. The operator T_t is very clear in this respect: not only does it symbolise the time evolution of the system itself, but the appearance of the time variable in the subscript indexes the operator which applies to each state to a given time. Of course, this does not yet entail a deep inconsistency with atemporal dynamics: as observed before, the iterative property of T_t and the potential for reducing arbitrary operators to T_{t_1} are, if anything, liable to generate the opposite impression, with one operator consistently responsible for evolving every state in the phase space trajectory. This subscript does not, then, mean that the operator is time-dependent in any deep sense. But the dynamical systems framework does still highlight weaknesses in aspects of the atemporality conjecture, such as the TVT—weaknesses consecrated not only in the narrowness of this thesis' understanding of dynamical systems, framed as differential equations rather than other types of functional relations, but also in its declaration that superficial criteria, such as the appearance of t or T in laws, disqualify them, or rather qualify their dynamics as being problematically time-dependent.

Within dynamical systems theory, a particular class of systems known as ergodic systems are of special interest, as well as possessing foundational significance in the context of statistical mechanics (for instance, in the characterisation of thermal equilibrium and systems' approach to thermal equilibrium). In fact, one of the important links between statistical mechanics and the temporality of determinacy can be envisioned through ergodicity. The next few paragraphs will therefore provide introductory material which should initiate the reader into this concept.

Whilst defining ergodicity precisely requires the conceptual apparatus of the triple $[X, \mu, T_t]$ to be expanded upon, the basic meaning of this property can be sketched out indicatively from the outset. As indicated previously, the measure μ translates the ordinary notion of size into the abstract context of dynamical systems. Ergodic systems are broadly those for which the time spent in some phase space region $A \in X$, which is to say, some subset of all dynamically possible states for the system, is proportional to the measure of A. Alternatively, to follow the presentation of Werndl and Frigg, ergodic systems are those for which the proportion of time spent in such a phase space region is equal to the measure of A.[49] More formally, consider the concept of a time-average

$$L_A = \lim_{t \to \infty} \frac{1}{t} \int_0^t X_A T_\tau (x) d\tau$$

where X_A is defined as 1 in the event that $x \in A$, but 0 otherwise. Roughly, this is a means of expressing more formally the intuitive notion of the amount of time a particular trajectory spends in A. The formal equivalent of the definition above becomes the claim that ergodic systems are those for which, for all measurable A and for almost all initial conditions, the following holds:[50]

$$L_A(x) = \mu(A)^1$$

How can this be spelt out more intuitively? Loosely speaking, ergodicity is often taken as in some way rigourizing the commonsensical notion of randomness. If the measure of the system's phase space is an abstract equivalent of the size of this region or of the relative amount of states in this region which can be occupied, a system which explores these states simply in proportion to this measure would appear to be, loosely speaking, *unconstrained* or indiscriminate in its behaviour and, in some sense, random. However, as will soon be seen, the relationship between ergodicity and random or indiscriminate behaviour is nuanced, and requires a number of more subtle properties belonging to the ergodic hierarchy to be distinguished.

The ergodic hierarchy must also be distinguished from ergodicity proper. The hierarchy refers to a classification of various different species of ergodic systems, each of which obeys a stricter set of dynamical properties than the last. Ergodicity itself is therefore only one property among many progressively more restrictive properties. Within ergodic systems, there are weak mixing systems; within those, strong mixing systems; within those, Kolmogorov systems; and within those, Bernoulli systems. Thus, all Bernoulli systems are ergodic; not all ergodic systems are Bernoulli. As this section will show below, mixing systems (and, by implication, the systems further up the ergodic hierarchy) display a characteristic key to understanding the impact of the theory of dynamical systems on the temporality of determinacy.

Before elaborating further, it is useful to adumbrate how ergodic characteristics such as mixing arise in the theoretical context of statistical mechanics, the field which frequently stimulates questions belonging to the metaphysics of time. As discussed in the previous section, one important such question concerns the relation of the microscopic components to the macroscopic observables of thermal systems, such as gases. The tendency for systems to equilibrate, for gases to spread uniformly or for heat to dissipate as can be observed in everyday circumstances, should be explicable by both microscopic and macroscopic approaches if they are to be at all reliable—by kinetic theory, which considers the underlying dynamics, as much as by thermodynamics, which focuses on phenomenological variables. Statistical mechanics is the discipline which attempts to reconcile these two different approaches, and in which ergodicity figures. By way of illustration, Boltzmannian statistical mechanics posits a distinction between macrostates and microstates, that is, states of the system which can be described in terms of macroscopic parameters such as temperature and pressure, and on the other hand their underlying constituents. The former supervene on the latter.[51] One of Boltzmann's insights was that vastly different numbers of microstates could correspond to distinct macrostates; in particular, equilibrium, high-entropy macrostates corresponded to far more microstates than far-from-equilibrium, low-entropy macrostates. Informally, there were more ways of arranging the constituents of a gas so that it is spread evenly throughout a container, more ways of arranging the constituents of two bodies in thermal contact, than were these systems in highly constrained positions, with the microscopic molecules or heat concentrated in one small region. Then, by assuming the phase space measure of the macrostate in question to be proportional to the probability of the system occupying that macrostate, the Boltzmannian approach could show that systems tended overwhelmingly to be found in higher-entropy or equilibrium macrostates. Crucially, the underpinning of this key assumption about proportionality, and its dynamical groundwork, lay in ergodicity; it accounts for the above relationship between the measure of a phase space region and the time spent in it.

This excursion through these basic principles provides a useful line of sight on the use of the basic notion of ergodicity within Boltzmannian statistical mechanics.[52] More importantly though, given this *précis* of the role of ergodic hierarchy in thermal physics is accepted, what bearing does it have on the goals of this chapter, namely to illustrate how certain features of members of this hierarchy impinge upon temporality? In Werndl's paper "What Are the New Implications of Chaos for Unpredictability?" she sets out an argument with this result. This argument links two properties: (i) the property of mixing, which distinguishes mixing systems which belong to the ergodic hierarchy, and (ii) probabilistic irrelevance between the states to which these events correspond.[53] These together undermine the prospect of atemporally determinative functional relations between the events.

First, consider (i). Like ergodicity itself, mixing systems can be given an intuitive gloss. Frigg, Berkovitz and Kronz consider the straightforward case (originally proposed by Gibbs) of a substance being poured into water, and define the overall concentration as the ratio of the volumes (or, abstractly, the measures μ) of the two substances.[54] Mixing in this case involves the time-evolution of the substance poured into water—assuming it is mixed or stirred suitably over time. In such a case, the concentration for any sub-volume of the combined fluid (however small) of the two becomes the overall concentration. In effect, the substance spreads out until distributed perfectly evenly. An initial ratio of one part poison to four parts water, after mixing, permeates the mixed substance so that for any chosen region within the mix the relative concentrations are one-to-four. A system is deemed to be mixing if and only if this state of affairs obtains given an infinite amount of time for the mixing to take place. Formally:

$$\lim_{t\to\infty} \mu\left(T_t X \cap Y\right) = \mu\left(Y\right)\mu\left(X\right)\left(\text{PI1}\right)$$

For two systems X, Y being mixed, an operator T_t governing time-evolution, and a measure μ.[55]

How does this definition (i) connect to (ii)—probabilistic irrelevance—and in what way does this connection enlighten us as to the temporality of the ergodic hierarchy? Werndl's work provides a useful demonstration of this. Werndl considers pairs of events represented by the two variables X, Y occurring respectively at a preceding time t_{-n} ($n > 0$) and later zeroth time t_0, and shows that the properties of mixing systems imply approximate probabilistic irrelevance in the following sense:

$$\lim_{n\to\infty} p\left(X^{t_0} | Y^{t_{-n}}\right) - p\left(X^{t_0}\right) = 0\left(\text{PI2}\right)$$

For all events X. The equation simply states that the difference between the probability of event X occurring at the relevant time *given A* occurring at t_{-n} and the simple, overall probability of event X occurring is zero, as n approaches infinity and the times of the two events become separated sufficiently. The term $p\left(X^{t_0} | Y^{t_{-n}}\right)$ is thus the conditional probability of X^{t_0} given $Y^{t_{-n}}$. The latter is probabilistically irrelevant to the former in the sense that the *conditional* probability is no different to the *pure* probability for sufficiently large n.

If this is the equation which grounds probabilistic irrelevance, how is it derived? Werndl's justification has several elements, but the most important is the equivalence she assumes between the probability of given state and the measure of this state, as a member of a state space.

This allows the mathematical properties of the measure μ defined on regions of phase space, and thus of states, to be naturally brought to bear on probabilities of these states at a given time. Then, by regarding events as states at associated with times, one can derive the more natural notion of event-based probability.[56] The measure of a state in phase space equates to the probability of that state occurring at a given time; the measure of an intersection of two states in phase space equates to their joint probability. From this equivalence, the second thesis of probabilistic irrelevance (PI2) can be derived from the first thesis of probabilistic irrelevance (PI1) using the following familiar definition from the foundations of probability theory:

$$P(A|B) = \frac{P(A \cap B)}{P(B)}$$

This is the foundation for the claim that mixing systems exhibit a form of probabilistic irrelevance, between the events or, strictly, phase space states which comprise them, subject to suitable assumptions. In her 2009 paper, as well as in her essay "Determinism,"[57] Werndl proceeds to use this proposal to associate the advent of chaos in scientific theories with these systems, and thus to answer the question as to what the sense in which chaos is unpredictable—in ways which other deterministic are not—might be.[58]

As mentioned above, mixing systems are only one species of ergodic systems. Probabilistic irrelevance is not only an attribute of mixing; indeed, it is extremised by the case of Bernoulli systems, the crown prince of the ergodic hierarchy, for which not only do no preceding events confer determinacy on later events; preceding events are all probabilistically irrelevant to later events, hence the comparison of the states of Bernoulli systems to the outcomes of ideal sequential trials, like the constituents of the eponymous probability distribution. What is noteworthy for the overarching purposes of this chapter, though, is that the status of mixing systems within the theoretical framework of modern science, such as the foundations of statistical mechanics, carves out a space in this domain for the temporality of determinacy. Aside from its implications for predictability, which Werndl quite explicitly classifies as an epistemological rather than ontological attribute,[59] the probabilistic irrelevance of events to one another—or even their absolute independence—vitiates descriptions of the time-evolution of systems. The prospect of laws which determine events at a given time as functions of determinate events at an arbitrary time cannot be accommodated for systems in which events are probabilistically unrelated given sufficient time. *A fortiori*, they cannot be accommodated for Bernoulli systems, where events acquire determinacy as they happen, and are probabilistically independent of other events in a strong sense: there is an order in which events are determined, but one identical to the order in which these events occur. Whereas, according to the classical vision of atemporal determinacy, a set of initial conditions coupled with specified dynamics sufficed to fix all outcomes and all antecedents, the framework of dynamical systems theory makes it clear that a specification of $A \in X$ coupled with knowledge of T_t is not always enough to determine the evolution of A into other phase space regions at other times. Werndl's perspicacious analysis of the properties of the ergodic hierarchy provides just one example of this insufficiency.

Revisiting the quadruple of atemporality theses in light of the contents of this section reinforces this conclusion. Contrary to the TVT, the relevant functional relations which provide probabilistic descriptions of events do feature occurrences of t, such as can be seen in the characteristic expression of probabilistic irrelevance PI2. The time-evolution operator of the dynamical system T separates events enough for their probabilities to become uncoupled. Contrary to the TVC, the least unnatural equivalent of the law it refers to in the context of ergodic dynamical systems, being the operator T_t evolving states along trajectories through phase space, changes truth values over time in a basic sense. As has been noted above, the iterative property of the operator in dynamical systems means that successive application of the original operator T_1 can generate all later transformations: $T_7(x) = T_1^7(x) = T_1T_6(x) = T_1T_1^6(x)$. Clearly, though, these relations are not preserved with variations in the index 1 as one would expect from the TVC. This basic observation shows how the immalleable framework of the atemporality conjecture fails to account for the alternative formalism of dynamical systems theory, even at the level of the schemata which appear in triples such as $[X, \mu, T_t]$. The TVC denies the role of time in laws possessing any dynamical significance. Further, when passing from T_t, considered as the equivalent of laws, to rules governing specific systems such as PI2, it is difficult to see how variation in t, whether $t_0 \rightarrow t_{-n}$ or $t_n \rightarrow t_m$ for $m \neq n$, holds any greater promise for the TVC. And thirdly, contrary to the thesis of Translation-Invariance, there are transformations such as these which do not map solutions of the dynamical laws, which is to say states $x \in X$ which evolve according to T_t, onto other solutions—for instance, in the transition between non-equilibrium and equilibrium states.

This concludes an outline of the sense which the framework of dynamical systems theory, frequently deployed in discussions of the foundations of statistical mechanics, and ergodic systems such as mixing systems, together provide for the temporality of determinacy. Such systems instantiate an important property—probabilistic irrelevance, taken by Werndl to define chaos and unpredictability—whilst remaining deterministic. This property applies to events at an appropriate temporal remove from one another whose conditional and unconditional probabilities of occurrence are, subject to reasonable assumptions, equal. This produces an awkward result for any program which seeks to cast laws in terms of time-independent or eternal functional relations, or which views the determinacy of events as being guaranteed by other events arbitrarily far into the relative future or past, as well as much of the accompanying content of the atemporality conjecture.

Aspects of Retrocausality in Quantum Mechanics

Thermodynamics is by no means the only area of modern science which hosts quandaries to do with time and determinism. The remarkable developments in the first half of the Twentieth Century in quantum mechanics, as well as the dedicated research into its philosophical foundations in the latter half, have led to a growing rhizome of metaphysical questions. These have culminated not only in competing interpretations of the now extremely well validated quantum formalism, but in contrasting predictions, evident in, for one, the famous inequality published by John Bell in 1964, as well as related no-go theorems such as the Kochen-Specker theorem or the Clauser-Horne-Shimony-Holt theorem. These results translate ostensibly superfluous or peripheral philosophical questions into testable hypotheses, whose verification or falsification impinge substantively upon the epistemic and ontological ramifications of the quantum theory. Such results have been accelerated in recent years with the development and adoption of an "ontological models"[60] framework enabling auxiliary philosophical hypotheses about the central object of quantum mechanics, the quantum state, to be formalised in the predictive language of the theory—further emphasising the continuity of the science itself with its metaphysical surroundings.

In literature on the topic, a privileged link is often taken to exist between quantum theory and indeterminism, indeterminacy, and related terms.[61] Why has this link been asserted? As to the first of these terms, Earman, in his *A Primer on Determinism* and "Determinism: What We Have Learned and What We Still Don't Know", challenges the association of quantum theory with indeterminism and entertains the reverse, pointing to a number of respects in which the theory is *more* deterministic than classical mechanics—the latter, he claims, often being lazily interpreted as an obvious paradigm of this feature. The central law of quantum mechanics, the dynamical equation known as the Schrödinger equation, transforms quantum states uniquely onto other quantum states. This equation expresses the evolution of these states, which inhabit an abstract vector space known as a Hilbert space, in terms of an operator, the Hamiltonian $\overline{H}$ and associated wavefunction φ: $\widehat{H}\varphi = i\hbar\frac{\partial\varphi}{\partial t}$. This is in effect taken to encode the transformation of these states over time by an evolution operator $\widehat{U}$ so that $\widehat{U}|\varphi_t\rangle = |\varphi_{t'}\rangle$. This equation is the nearest equivalent of the Russellian functional relation *f* which fixes or determines quantum states in terms of other quantum states at distant times, and in this sense fulfils the schema in earlier chapters:

> A system is deterministic if, and only if, for events $e_1, e_2, \ldots, e_n$ occurring at times $t_1, t_2, \ldots, t_n$ and for any state E_t of this system at any time t, there exists a function relating the $e_1, e_2, \ldots, e_n$, $E_t = f(e_1, t_1, e_2, t_2, \ldots e_n, t_n, t)$.[62]

As well as these properties of the Schrödinger equation, Earman adduces other morsels of evidence for determinism in the quantum domain. He points to the fact that the sensitive dependence on initial conditions present in classical mechanics, such that phase space states separated by small perturbations can follow wildly divergent trajectories, does not seem to exist in quantum dynamics: states which at a given point in time begin to evolve in each other's neighbourhood remain so for all future times.[63] Nor, he denies, are the puzzling discontinuities of classical dynamical models present in the trajectories of quantum systems—discontinuities such as those of the inelastic, impenetrable molecules conceived by these models to interact instantaneously, and which generated alternative systems of natural philosophy to the Newtonian view, such as that of Boscovich.[64]

As Earman himself observes, however, a distinction must be drawn between the decision to regard quantum theory as deterministic or indeterministic, and other decisions as to the determinacy, or determinateness, of the states it describes. From the claim that the functional relations of quantum theory do not determine whether a future outcome will or will not obtain it does not follow that it is indeterminate whether such an outcome will obtain, even if it does follow that such an outcome is undetermined.[65] Indeterminate states may evolve in accordance with laws, no less than determinate states can evolve indeterministically. So what of the question as to determinacy in quantum mechanics? What implications, if any, do the many discussions of quantum mechanical determinism or indeterminism, Earman's among them, possess for the latter? One clue lies in the importance of *superposition* in foundational presentations of the discipline. Superposition refers to the decomposition of the state describing any quantum system into a linear combination of other states, typically known as eigenstates or basis states. These states are often held to stand in a privileged relationship to quantum mechanical operators,[66] which, under certain conditions, represent measurements of variables of interest such as the momentum or spin of a particle.

Rightly or wrongly, many authors have taken superposition as being the crucial representation within the formalism of indeterminacy, as the following examples suggest.[67] Consider first David Albert who, in his succinct introduction to quantum problems *Quantum Mechanics and*

Experience, uses superposition to introduce some of the theory's time-honoured philosophical quandaries, considering a fictional property of an electron, whiteness, and its possible different locations. Albert envisages an electron which is initially white travelling through a particular measurement apparatus. This apparatus admits of two possible states which can be occupied, which are each positions, on *s* or on *h*. One might assume that the electron travelling through the apparatus is either on *s*, or on *h*, but not both, nor neither; the two are disjoint, non-overlapping, but the electron must be in one. For Albert, though, of the electron one can say only that it is neither on *h* nor on *s*, nor on both *h* and *s*, nor on neither of the two, but rather in a superposition of the two states. He points to the clear interpretative difficulties posed by such a state:[68] a system (including a particle) in a superposition of different states, which might take the form $\alpha|h\rangle + \beta|s\rangle$, is in neither state determinately. For those willing to convert a negation of determinacy into indeterminacy, this indicates an indeterminate state.

Many other examples of authors drawing this connection exist. By way of a second case, consider how, in Dirac's discussions of the measurement process, such as the analysis "Superposition in determinacy" in his 1930 tract *Principles of Quantum Mechanics*, a necessary and sufficient link is asserted between superposition and indeterminacy. He claims quite directly that relationships of superposition between quantum states implies the existence of indeterminacy.[69] Relatedly, Barrett, in his *The Quantum Mechanics of Minds and Worlds*, refers to Dirac's notion of superposition as one in which the quantum state in fact possesses what he describes as properties *intermediate* between two states.[70] Jessica Wilson, in her "Determinable-Based Account of Metaphysical Indeterminacy", associates quantum superposition with what she characterises as an undetermined determinable. A superposition state consists of eigenstates of operators, each of which is associated with a measurable property (of momentum, energy, spin, etc.).[71] Since, she explains, systems possess determinate properties (of momentum, energy, spin, etc.) if and only if they lie in an eigenstate of an operator so associated, and since superpositions fail this criterion, they consist of undetermined determinables rather than determinate states. Wilson argues that these undetermined determinables are in turn related to metaphysical indeterminacy, and provide an empirical basis for the latter.[72] E. J. Lowe, in his "Vague Identity and Quantum Indeterminacy," refers to the quantum mechanics of particle-interaction as evidence for propositions which are indeterminate ontically, rather than semantically based on superposition and entanglement.[73] Moreover, these ontic implications of superposition, entanglement and related quantum phenomena are not the only ones: in Jeffrey Bub's semantic analysis, the indeterminacy of quantum systems is represented in the fact that quantum state may or may not exhibit certain properties. For such systems, well formed propositions exist for which no determinate truth value is assigned for that state.[74] This limited collection of examples provides some evidence for the close relationship posited by scholars between quantum systems in superposition—that is, systems whose states decompose into linear combinations of basis states—and indeterminacy.

This quick survey provides some context for the meaning of indeterminacy often considered to be latent within quantum systems, encompassing both ontic and semantic variants, and often tied to the principle of superposition. The vast difficulties confronting any attempt to interpret the quantum state are clearly visible in this principle. One can intuitively grasp the meaning of the variables of classical physics, such as the x or p of position or momentum. One can interpret these symbols as representing an unknown quantity or one which varies over time. But, in the quantum domain, as Albert shows, it is rather hard to discern the physical meaning of combinations of vectors and complex numbers, or to connect them to a state. The challenge of interpreting the quantum state in a way consistent with the predictions of quantum theory is a defining one for theorists working in the foundations of the discipline.

In order to see how this impinges upon the questions of temporality, this problem must be extended somewhat. Whereas questions as to the determinate or indeterminate nature of the quantum state have for the most part to do with the nature of the quantum state in itself, it is the *evolution* of this state which is placed in question by the temporality of determinacy. More precisely, a question or class of questions often subsumed under the rubric "the measurement problem" in quantum mechanics point to the discord between two aspects of the theory. On the one hand, the quantum state consists of superpositions. The dynamics of the theory, captured in the Schrödinger equation, typically preserve such states: superpositions remain in superposition over time. More specifically, superpositions remain in superpositions even when they interact with other systems. On the plausible assumption that quantum mechanics is a complete theory, its predictions should hold not only for the system, but also for others with which it interacts, including its surroundings, as well as the apparatus used to observe and measure the system. It seems to follow that the quantum system exists in superposition not only when it evolves undetected, left to its own devices, but that measurement outcomes, more properly described as combinations of the system of interest with some experimental apparatus, are themselves in superposition.[75] Thus, according to the quantum mechanical laws of nature, measurement of a system in superposition itself leads to superposition—this latter superposition combining both the measurement apparatus and the system being measured.

This observation compounds further the interpretative puzzles posed by the quantum state. Even attempts to measure systems in superposition generate systems in superposition. Yet, we are able to observe determinate measurement outcomes when performing experiments. How can these two be reconciled? A valid theory must be consistent with measurement outcomes. But whilst superpositions seem incompatible with experimental outcomes, not to mention everyday observations, the only way of describing a quantum state in a way consistent with the now overwhelming preponderance of quantum phenomena, from entanglement to interference, is the vector formalism above, which predicts the preservation of superposition over time. This antagonism has stimulated a range of responses from theorists, responses which effectively *define* many of the competing interpretations of quantum theory. Taking a stance on the "measurement problem" in many cases means adopting an interpretation of the quantum state, and each of these interpretations harbours different implications for the temporality of determinacy. The next section surveys a selection of these.

A natural starting point would be traditional, orthodox, or Copenhagen quantum mechanics—which is in fact a diffuse collection of approaches. These supplement the dynamical law of the theory (the Schrödinger equation) with a rule known as the projection postulate, or collapse principle. Spontaneously and stochastically, the superposition "collapses" into a determinate eigenstate when measurements occur. In Dirac's terms, measurements cause a given system to suddenly move into an eigenstate of the variable being measured, the eigenvalue of this eigenstate being equal to the measurement outcome.[76] Such an interpretation therefore denies that the Schrödinger equation continues to evolve the system in superposition for the specific circumstance in which a measurement is performed. Clearly, this limitation on the scope of the dynamics of the theory to all situations except those in which measurements are being performed has an air of arbitrariness, and poses the difficulty as to exactly what "measurement" means. Bell, in his paper "Against measurement," provides an extended list of terms he deems to fall foul of this difficulty.[77] If quantum theory is to be complete, and anthropocentrism is to be avoided, *any* interaction between *any* two systems could be regarded as a measurement, and not merely interventions of a privileged human observer and the apparatus they wield. Indeed, as Wallace notes, the principle of collapse is arguably foreign to the practice of experimental quantum mechanics, and does not apply at all straightforwardly to cases such as unsharp or continuous measurements.[78] Yet it is this principle which is so frequently espoused as standard by commentators, such as Barrett, who claims that it must explain the fact that our everyday experience is determinate by guaranteeing that determinate physical records exist at once measurements have been performed.[79, 80]

The collapse principle, in the stochastic form in which it is often held out as orthodoxy, temporalizes the determination of the quantum system. An exogenous intervention, not subject to the Schrödinger equation, is required to transform superposition into determinate state. The determinacy of the system which is acquired at the point of collapse is thus no consequence of the functional relation ($\hat{H}\varphi = i\frac{\partial \varphi}{\partial t}$) which preserves superposition over time. Crudely speaking, if superpositions comprise indeterminate states, and collapses of superpositions determinate ones, collapse is the instantaneous process of determination. An *ex machina* influence unaccounted for within the Schrödinger dynamics, undetermined and undeterminable until the last moment, comes from without to precipitate collapse. The collapse law in this sense introduces the temporality of determinacy into the evolution of quantum systems.

This putative orthodoxy is not the only interpretation of quantum theory; nor is it even the only interpretation which brings collapse into its core hypotheses. Dynamical collapse theories, most notably the Ghirardi-Rimini-Weber (GRW) theory, attempt to provide law-like account of the collapse process, explaining what induces collapse in a more transparent way than leaving it to the inscrutable interventions of an observer, apparatus, experiment or similar. Collapse in this framework becomes a mechanism belonging to the theory's dynamics, rather than a supplemental fudge. In the GRW case, this mechanism takes the form of spontaneous localisations which occur randomly to constituents, and with a frequency which increases rapidly with the size of the system. For macroscopic objects such as those directly observable by experimenters, the GRW probability becomes overwhelming that such objects are found to have collapsed, promising to reconcile quantum theory with our observations of determinate states. GRW provides a mathematical model of where and when in an object localizations occur. Aside from this, other collapse theories, such as that of Roger Penrose, provide explanations for collapse mechanisms in terms of quantum gravity.[81]

From the standpoint of these approaches to the evolution of the quantum state, the temporality of determinacy takes a slightly different form when compared with supposedly orthodox conceptions of collapse, couched in terms of observation or measurement. Dynamical collapse theories can specify detailed functional relations which govern the localization process. These describe how a system's state is determined and becomes macroscopic. Nonetheless, it remains a time-dependent stochastic process: the functions, in conjunction with a specification of a state, need not fix future states, but rather specify a probability distribution which they must obey, depending on a frequency parameter.[82]

These are far from the only responses to the problems in quantum theory which, as this section mentions, are in part the very inspiration for its different interpretations. Among the first alternatives to emerge from the milieu were hidden variable theories, which hold the quantum state to be incomplete descriptions of reality, among them the famous De Broglie-Bohm theory. For these theories, measurement outcomes are determined by defined observables which are not visible in the quantum formalism. These often accord epistemic rather than ontic import to the existence of superposition, meaning the linear combination of states comprising the quantum state reflects our ignorance about which feature the world really has, and not so much some intractable feature of the world which requires the classical manifest image of nature to be revised wholesale: it thus, to some degree, protects our intuition that systems' properties must possess determinate, sharp or well defined values at all times. However, challenges to these alternatives by historical results such as Bell's Theorem[83] and the Kochen-Specker theorem, as well as recent results building on these, such as that of Pusey, Barrett and Rudolph,[84] have undermined the possibility of formulating them without either abandoning intuitive assumptions such as locality and contextuality, or miring them in tension with the predictions of quantum theory.

However, recent years have seen significant momentum build behind the Everettian approach to the quantum state, often glossed as the many-worlds theory. Hugh Everett's 1957 statement of this approach has been revitalized by philosophers such as David Wallace and Simon Saunders. Everettian quantum mechanics takes the quantum state as a complete description of reality, and superposition to describe real, determinate, physical worlds which, subject to certain conditions, evolve independently of one another: rather than linear combinations of states used to describe quantum systems representing our ignorance about which of them is singularly determinate or real, or the failure of any to be determinate or real, such combinations describe a multiplicity of determinate, real states. Thus, in an ironic parsimony, adherents to the Everettian approach claim not so much to expand the ontological commitments of the theory, but to recapitulate quantum mechanics itself, unadulterated and shorn of interpretative overlay, following Bryce DeWitt's entreaty that the predictive formalism be allowed to yield its own interpretation.[85] Contrary to imputations of a bloated ontology, then, Wallace claims to advance a conservative interpretation of quantum theory, on the basis that it in no way modifies the mathematical formalism.[86] Since the dynamics of the Schrödinger equation are all there is to the Everettian approach, it is deterministic by nature, and since it links each state, whether or not superposed with others, to a real state of affairs, it also allows the classical vision of the time-independent laws of nature, in the form of functions which capture the theory's dynamics, eternally fixing the world's evolution, to be rehabilitated in the quantum domain—*prima facie*.

This cursory survey of the basic dynamics of quantum theory, together with a small sample of competing interpretations of the quantum state, shows that the status of the temporality of determinacy in this field is interpretation-dependent. Indeed, an account of the relations between concepts such as determinacy and determinism and quantum mechanics would fill and overspill many volumes, never mind a brief review. The purpose of this survey is therefore only to establish the following modest fact. It is no more obvious from the *fundamentals* of the theory itself whether the workings of its laws display marks of time-dependency than it is for classical or macroscopic systems. The question turns on a number of decisions, as to the ontic or epistemic characterisation of the state, the completeness or incompleteness of the Schrödinger dynamics, and the deterministic or stochastic nature of the system's evolution. So does it follow from this that there is nothing determinate to be learned from the theory as to the temporality of determinacy? If the impact on the latter depends on which of sundry competing interpretation one adopts, must any analysis of this question be futile until such competition is resolved? In spite of the divergences between quantum schools of thought, there are sub-fields of the discipline, many of them burgeoning in the last two decades, which provide fertile material for an investigation into the temporality of determinacy—material which does not just repeat the traditional debates between the traditional interpretations. The remainder of this section aims to present a selection of these sub-fields, all of which relate to the notion of retrocausality, focusing particularly on the discussions of the last two decades, and to relate them to the perspectives on physical laws considered in this text.

A suitable starting point is the work of Huw Price, who has become one of the most innovative recent contributors to discussions of the role of retrocausality in quantum mechanics, in articles such as "Does Time-Symmetry Imply Retrocausality? How the Quantum World Says "Maybe"" and "Toy Models for Retrocausality" as well as the more accessible "Dispelling the Quantum Spooks—a Clue that Einstein Missed?"[87] Price's argument that retrocausality is a commitment entailed by some plausible assumptions quantum theory might well be thought to contain is labelled the time-symmetry argument for retrocausality (TSAR). It proceeds according to the following schema:[88]

Premise (1): Realism
Premise (2): Time-symmetry
Premise (3): Discreteness
Conclusion (C): Retrocausality

Price imagines two cubes, each of which contains ports equivalently positioned on perpendicular faces, capable of emitting and absorbing photons. The ports are oriented so that the paths of the photons emitted coincide at each cube's centroid. In turn, within each cube, a beam-splitter is situated at a variable angle to the path of the two photons. The two cubes are connected by a beam of photons emerging from one box and incident on the other. This setup thus opens up a path from either of two entry points, through a beam splitter, across a vacuum, through another beam splitter, and onto either of two exit points—in two possible directions, given the symmetry of the apparatus. For any of the two boxes in this setup, a probabilistic function relates three important variables: (a) the polarisation of the photon beam leading towards the other box (τ_L, for the case in which the photon travels from left to right), (b) variation in the angle of the beam-splitter (σ_L or σ_R, depending on the box) and (c) variation in the port through which the photon enters into the box, or exits from it.

After outlining the setup, Price provides definitions of the metaphysical assumptions (1)–(3) underpinning his demonstration. Realism he takes to entail broadly that the polarization τ_L corresponds to a real property or element of reality, citing Bell's definition in his well known discussion "Beables for quantum field theory." Time-symmetry he takes to mean that, for any physically possible process, that is, for any set of solutions of the relevant dynamical equations, together with the constraints specific to the experimental setup, the transformation $t \rightarrow -t$ maps such solutions to another set of solutions. More heuristically, if a process were observed first in one temporal direction and then in another, there would be no way of telling which of two directions could be associated with the forward or positive direction, nor which came before or after in time.[89] Finally, Price assumes discreteness. This assumption restricts the range of inputs to, and outputs from, each box to discrete values: each photon entered into the apparatus is entered either into one port or the other; superpositions are debarred. Pusey and Leifer provide a minimal definition of Price's assumption and its purposes: should a photon be detected at an output port of a beamsplitter, no photon emerges from the other ports of the beamsplitter—a necessary condition to ensure that the photon leaving the beamsplitter can be considered as the reverse in time of a photon being inserted into the beamsplitter through a certain port.[90]

How does this experimental setup, coupled with the assumptions (1)–(3), generate an argument for (C)? First of all, Price notes that, assuming discreteness, no superpositions of quanta can be entered into the left hand side of the device. It follows from this that a suitable choice of the orientation of the beamsplitter in the left box restricts the possible range of orientations τ_L independent of the entry point of quanta entered. The orientation choice σ_L yields possible output polarizations limited only to σ_L and $\sigma_L + \frac{\pi}{2}$. Second of all, Price notes that, assuming realism, these output polarisations τ_L are real or, again in Bell's terms, are beables, rather than abstract or unphysical observables.[91] Third of all, Price notes that, assuming time-symmetry, the relations between τ_L, σ_L and the ports through which photons enter or exit the left hand box must be equivalent to the relations between an equivalent set of observables τ_R, σ_R.and the ports of the right hand box. Otherwise, a symmetric arrangement of the experiment (in which the two boxes, with their ports, beamsplitters and polarized beam) would be distinguishable under the transformation $t \rightarrow -t$. After all, τ_L, τ_R are variables which simply capture the relationship between changes to the left hand setup to and changes to the right hand setup. However, it follows from the conjunction of these three assumptions that the choice of orientation σ_R influences τ_R in the same sense in which the choice of orientation σ_L influences τ_L. The latter, by virtue of realism, is a real quantity. Thus, to the extent that the experimental *controls* or *effects* τ_L by their decisions about σ_L the same holds of τ_R by their decisions about σ_R.

Does this establish that retrocausality is part of this framing of quantum mechanics, in a meaningful sense? Assessing this question requires the loaded and connotation-ridden term retrocausality to be unpacked, defined more precisely, and separated from the readily conjured philosophical clichés of time-travel and bilking paradoxes—ones hamstrung by causal models poorly suited to the quantum mechanical domain. In his various writings on the topic, Price's remarks on the subject are diverse, and in his various expositions of his argument for retrocausality—or, perhaps, his argument for the validity of this argument, cf. its soundness—provides a number of different perspectives on this question. In his paper with Ken Wharton "Dispelling the Quantum Spooks—a Clue that Einstein Missed?" retrocausality is discussed in terms of *control* exercised over some real state or beable occurring prior to the real state or beable in the hands of the measurer, or the *influence* so exercised by the latter.[92] Price's intuition appears to be that if there is a sense in which a future state which is a function of an observer's intervention in an experiment is caused by such an intervention, so too can a past state be caused by a similar intervention, hence the prefix "retro" in the description "retrocausality" he employs. The properties of a system prior to a measurement are thus taken to be under the control of an experimenter who chooses a setting for the measurement in question. The terms "future" and "past" here, as applied to events, may mean as little as being indexed by a time value greater or lesser than others and need imply no decision with respect to the reality of different modes of time. Retrocasuality in this sense simply means the ability to control past beables, and thus past events. Elsewhere, such as within his *Synthese* paper "The Philosophy and Physics of Affecting the Past" Price makes a number of looser references, such as to effects occurring earlier than causes by way of evidence of backward causation and backward influence.[93] Similar terms appear in his extended work *Time's Arrow and Archimedes' Point: New Directions for the Physics of Time*.[94] Separately, in his main exposition of his argument for retrocausality, "Does Time-Symmetry Imply Retrocausality? How the Quantum World Says "Maybe"," he supplies a counterfactual account of retrocausality, claiming of the right hand box that, with respect to an experimenter selecting the orientation of the beamsplitter, were $\sigma_R + \rho$ to have been the choice rather than σ_R, a different value of τ_R would have ensued.[95]

The impression that Price's argument involves some sleight of hand may linger here—as well as its vulnerability to obvious objections. It may strike the reader as odd that correlations between a measurement made at a time and others made prior to it are beatified with the moniker "retrocausal." Are these any stranger, any richer with metaphysical import, than the discovery of correlations between present observations and future outcomes? What distinguishes this argument from the argument that a functional relation of Russell's form $E_t = f(e_1, t_1, e_2, t_2, \ldots e_n, t_n, t)$,[96] relating as it does events at earlier and later times, evidences *retrocausal* relations between them in the same sense that it evidences causal relations? Surely no one would conclude that phenomena such as the Big Bang and the universe's early expansion, inferred from the meticulous observations of modern astronomers, arose as the retroactive effects of the present day. Is Price's argument any more plausible than these caricatures? The factor which separates Price's approach from them is its anthropocentrism: it is, with his account, only those relationships between events involving the intervention of an observer—relationships which therefore accommodate the notions of control and influence Price utilises—which qualify as candidates for retrocausality. General relations of the form $E_t = f(e_1, t_1, e_2, t_2, \ldots e_n, t_n, t)$ do not amount to counterexamples to Price's argument, unless the events involved actually exert a meaningful influence over one another. However, to the extent that this strategy defends this argument it leads to an unwelcome mystification. Precisely what sense can be given to the notions of influence and control, if this is the underpinning of retrocausality? What sense can be made of these notions in any account of quantum theory which is complete, and thus which accords no privilege to the human observer which is not extended to other physical systems, including being subject to the theory itself?

Recent commentary in the field provides a helpful answer to this problem. In a recent paper, Leifer and Pusey rigourize Price's argument, providing a clearer view of the meaning of retrocausality, as well as relaxing some of his critical assumptions and replacing them with more general ones, in order to broaden its scope. Simultaneously, they refine concepts such as time-symmetry, introducing a distinction between ontological and operational senses of the term.[97] One of Price's controversial premises in his argument, they swap out for a weaker principle they describe as λ-mediation: the physical properties of a system mediate any correlations between preparations of the system (the orientation of the beamsplitters in Price's example) and measurements subsequently performed on the system.[98] Consider the following statement the authors make of a no-retrocausality principle, i.e. one whose violation would imply the existence of retrocausal phenomena.

$$p(a,b,\lambda|x,y) = p(b|a,\lambda,x,y)\,p(\lambda|a,x)\,p(a|x).$$
$$(\text{No Retrocausality Thesis NRT})$$

Here, a corresponds to the uncontrolled *output* of what Leifer and Pusey describe as the preparation stage of the experiment (the input port of the left hand box through which the photon entered), b the output of the *measurement* stage (ascertaining whether the particle concerned exited this or that slit, perhaps with a pair of detectors), and x and y respectively the inputs to these two stages. Concomitantly, the variable λ labels physical properties of the system, the counterpart in this formulation of Price's reference to beables such as τ_L and τ_R. Each experimental run involves determinate values of input variables x and y and output variables a and b, yielding a determinate value for λ. The latter therefore represents properties of the system which the observer may not be able to observe.[99]

Recalling that $p(\alpha|\beta)$ represents the probability of α occurring given β occurs, the NRT thus concerns the conditional probabilities of different combinations of these inputs and outputs. $p(a, b, \lambda| x, y)$ is the conditional probability of the outputs of preparation and measurement and the physical state of the system based on inputs x and y. For no retrocausality, such a probability should be factorable into a product of several independent probabilities: the probability of ultimate output b (detecting photons exiting the right hand box at either point), the probability that the physical state λ obtains, and the probability of preparation input a. Most importantly, the probability of λ and a should be independent of outcomes occurring *later* in the process: neither the measurement output b, for instance, nor the measurement input y, appear in the argument of the expression for the probability of either. Strikingly, from the conjunction of this assumption with formal analogues of time-symmetry and λ-mediation assumptions, Leifer and Pusey derive a result, reproduced below, which they claim is incompatible with results in the domain of no-signalling quantum theory.[100]

$$p(a,b,\lambda|x,y) = p(a|x,\lambda)\,p(b|y,\lambda)\,p(\lambda)$$

Given that these results are veridical, and that the authors' argument is valid, one of the assumptions responsible for this equation must be abandoned. Leifer and Pusey in no way asseverate on this basis that the NRT is straightforwardly false; any of the others in the argument could be singled out. But given the intuitive plausibility of time-symmetry and the authors' beliefs about its fundamentality, and the overall relatively uncontroversial status of λ-mediation, this principle seems difficult to uphold.

It is obvious that the NRT thesis, and its negation, are rather different and rather more precise than the looser notions employed in other discussions of retrocausality, couched as they are in often perplexing notions of control, influence and movement backwards in time. As Leifer and Pusey observe, the assumptions made about which variables are under the control of an observer frame their apparent asymmetries. Violation of no retrocausality here means that variables exist of which the probabilities of output results occurring earlier in time are not independent. More particularly, $p(\lambda|\, a, x, y) = p(\lambda|\, a, x)$ is violated for a class of experiments that the authors envisage.[101] Price's argument is a special case of this more general principle, subject to stronger (more controversial) assumptions. What is more, whereas Price's account of retrocausality may have seemed difficult to distinguish from ordinary retrodiction through atemporal functional relations, the relevance of anthropic notions such as control and influence are here given a much more precise sense: the probabilistic relevance of events at the apparatus' outputs, represented by Leifer and Pusey by the variable y, to the real state of the system preceding these events in time. It thus becomes possible to judge whether or not retrocausality obtains by considering these correlations between individual events, rather than evaluating the status of a general law between all events, that is, the general function f which appears in Russell's definition of determinism. These advantages of increased generality and plausibility, as well as a number of other innovations, distinguish this reformulation of Price's argument.

These considerations clarify the implications of the work of Leifer, Pusey, Price and others for the temporality of determinacy. The probabilistic relevance or irrelevance of two non-simultaneous outcomes to one another once again grounds a distinction between the order in which the measurement outcomes or events occur and the order in which they are determined. In the context of Price's picturesque orchestration, unquestionably, when the orientation of the right hand beamsplitter is set is *later* than when the polarized beam enters the right hand box, for the case in which the experiment is begun with the entry of a photon into the left hand box. The polarization of the incident beam is an event in time, the orientation another; the latter occurs after the former. Yet, the orientation determines the polarization: expressing the probability of the polarization taking a certain value, a parameter included within the state of the world λ, independent of the beamsplitter orientation, leads to predictions which are violated by quantum theoretic tests: such is the conclusion of Leifer and Pusey with their reformulation of Price's argument and their exposition of the NRT. The non-monotonic relationship between these different events consists in the failure of the equality $p(\lambda | a, x, y) = p(\lambda | a, x)$, in spite of the fact that $t_y > t_\lambda$.

Leifer and Pusey are not the only recent scholars to unearth connections between probabilistic relationships and the metaphysics of time. E. Adlam's 2017 paper "Spooky Action at a Temporal Distance," a title which reanimates a 1947 witticism of Einstein's, explores the possible role of temporal nonlocality in the foundations of contemporary physics—and thus not only the *spatial* nonlocality often brought up as a response to such phenomena as entanglement. Adlam considers a variable $\lambda(t)$ which represents the state of the world as a function of time, measurement outcomes A and B, and measurement settings a and b, defining a principle of temporal locality in the following way:[102]

$$p\left(A,B \middle| a,b,\lambda\left(t_a\right),\lambda\left(t_b\right)\right) = p\left(A \middle| a,\lambda\left(t_a\right)\right) p\left(B \middle| b,\lambda\left(t_b\right)\right)^{cii}$$

Here, the decomposition of a joint probability of measurement outcomes into a product of individual probabilities, each a function only of states in the immediate temporal neighbourhood of these measurements, distinguishes this notion of locality—the intuition being that states of the world at later times, as well as spatially distant measurements, are probabilistically irrelevant to a given outcome. In both this case and the case of Leifer and Pusey, probabilistic correlations between measurement outcomes or, in Russellian terms, events, serves as a useful tool to pick out those groups of outcomes or events which confer determinacy on one another, beyond the mere requirement that they be relatable by a function. Events appearing together in the one function may qualify the system as a whole as deterministic, but probabilistic relations between some of these events and not others provides evidence that specific events within a system impart determinacy to one another in a stronger sense than does their appearance in one function. The occurrence of some events and not others, in conjunction with the probabilistic equations which govern them, enable the prediction and explanation of measurement outcomes characteristic of laws of nature. Contrary to the atemporality theses concerning these functions, which take their existence and structure to represent laws operating independent of time, the nature of the probabilistic correlations between the events they describe picks out groups of events probabilistically relevant or irrelevant to one another. In this case, the order in which these events *occur* is one thing; the order in which they are determined or, more exactly, the relations of probabilistic relevance or irrelevance between them, is another. The two need not correspond to one another, as Price's example, together with its derivatives, shows. Further—in the limited, precise sense that conferring determinacy on events is an essential aspect of a physical law—the prospect of events at specific past and *future* times taking on a privileged role in determining a given state indicates that the probabilistic relations concerned amount to manifestations of time-dependent laws.

Studies of the role of retrocausality in quantum mechanics occupy the most important terrain for an investigation into the discipline's implications for the intersection of determinism and time. But the hypothetical experiments of Price, Leifer, Pusey and others are not the only ways in which retrocausality crops up. For the purposes of this section, these are certainly worthy of exploration. The beginning of this chapter cited a number of John Wheeler's remarks in which the implications of retrocausal influences are dramatically rendered, and bound up with his references to present measurement choices and decisions reaching backwards through time, and to the greater part of the past being unformed or entirely non-existent until created. Wheeler's now famous dramatization of these claims, now adapted into Wheeler delayed choice or delayed choice quantum eraser experiments, take a particularly striking form. A beam of light is directed towards a detector screen, with a shield designed to block the light in its way; two holes or slits are punctuated in the shield allowing (putatively) a limited number of photons through. Rather than two separate piles of photons building up on the screen behind the slits, a continuous series of interference fringes, with a quasi-sinusoidal pattern, show up—even when the intensity of the beam is diminished to the point that photons reach the screen only once every hour, guaranteeing only one falls within the experiment at once. These evidence interference effects between different components in the superposition state which represents the progression of the photon through the experiment. It seemingly defies the expectation that the photon passes through one slit or the other. Bizarrely, though, this interference pattern is destroyed whenever a detector is inserted next to one slit—every time a photon is directed at the screen—suggesting an elliptical measurement dependency or contextuality to the constitutive variables of the experiment: the photon passes through one or another slit if one measures whether it does; it remains in superposition otherwise.

The "delayed choice" version of this experiment extrapolates this seeming measurement dependency out through time, up to the cosmological scale. By placing crystals on the side closest to the screen of both slits, pairs of identical, entangled photons take the place of any photon passing through any slit. Of each pair, the use of a prism means that one photon propagates towards the screen unencumbered, whilst another is projected away over a great distance (one far greater than the distance from the crystal to the original screen). At the end of this distance, a complex arrangement of beamsplitters is arranged which either keeps the two beams emerging from either crystal apart, so that any photon detected in either beam can only have passed through a particular slit, or recombines the two distinct beams so that distinguishing the provenance of any photon or series of photons becomes impossible. Thus, subtle changes to the state of the beams allowed to propagate away from the original slits and detector (such as their recombination) seemingly determines the outcome of experiments completed ostensibly much earlier. If the recombined beam is detected, the interference pattern is detected on the screen; if any of the separate beams are detected, the pattern is destroyed. The further the entangled pairs are allowed to propagate apart, the greater the positive temporal gulf between putatively completed experiment and the outcome which in fact, and despite occurring later, determines it.[103] This experiment concretizes the many idiomatic formulations of Wheeler, painting as they do a picture of an incomplete past subject to construction and reconstruction by agents' present actions.

These aspects of retrocausality in quantum mechanics are the primary sources in the theory of a distinction between the order of events and the order in which they are determined. From previous chapters' discussions—from, for example, the revisability of the value of financial securities to the retroactive confirmation of adversaries' intentions in military conflicts—this is a now familiar concept that this work has attempted to give to the temporality of determinacy. Rather than all events being determined independent of time, determination is a process; rather than this process being the passing of time, and preceding events determining their successors, influences can be exerted on events occurring previously in time. Such influences are characterised by probabilistic relations which obtain between the events in question, and need not entail a commitment to objectionable anthropocentrism. It is important to note, however, that in spite of the idiosyncratic nature of these thought experiments and their seeming esotericism, the history of quantum theory sets something of a precedent with respect to retrocausality, as this concept also arises in several other contexts.

The first of these which is particularly worthy of mention is its role in the transactional interpretation of the quantum theory. Although developed most fully in the 1980s, this interpretation has origins in the thinking of Feynman and Wheeler and, even earlier, in the analysis of the Einstein-Podolsky-Rosen experiment by the physicist Costa de Beauregard. Feynman and Wheeler's approach arose first as a solution to the problem of electron damping, in which the emission of radiation by an accelerating electron leads to a damping effect on its motion in a way inexplicable by the classical theory of electromagnetism.[104] In order to account for this puzzle, Feynman and Wheeler postulated an exchange occurring between this moving radiation source and the body which eventually absorbs the radiation, and consisting of both advanced and retarded waves travelling between the two. Costa de Beauregard, for his part, weaponised the distinction between advanced and retarded waves in an attempt to identify an unavowed assumption in the Einstein-Podolsky-Rosen thought experiment, in which entangled particles with related spin properties are allowed to propagate far apart, and in which a measurement performed on one yields a determinate outcome also for its distant partner. De Beauregard considered the possibility that an advanced wave propagating from the final measurement to the initial state of the system predetermined the spin property possessed by each of the particle pair.

Referring to this backdrop and to the Feynman-Wheeler approach explicitly in his presentation of transactional quantum mechanics, John Cramer defines the transactions which he takes to comprise the dynamics of the quantum world as exchanges of energy between emitter and absorber particles, mediated by two types of waves—an "advanced" or "confirmation" wave propagating backwards from absorber to emitter and a "retarded" or "offer" wave propagating in the opposite direction.[105] The symmetries of the exchange preclude oddities such as negative energy emission or backwards-in-time signalling. The wavefunction, according to this view, has physical reality as a forwards-propagating wave aimed at an absorber. The collapsed wave, conventionally taken to correspond to a determinate state, results only from the "transaction" or interaction between forward and backward propagating waves. Further, Cramer postulates a further correspondence which enables his physical vision of interacting waves to reproduce important elements of the predictive formalism of quantum mechanics: by envisaging the complex-valued wavefunction φ to correspond to an offer wave, and its complex conjugate φ^* to correspond to a confirmation wave, Cramer specifies a physical correlate of the notion of probability amplitudes and claims to derive one aspect of the Born probability rule for the probabilities of different quantum measurement outcomes.[106]

Cramer views the scope of his theory to extend to *all* quantum mechanical processes and to yield a complete, alternative interpretation of the quantum formalism. This he summarises in the form of five core premises: the uncertainty principle, the Born probability rule, the amenability of *all* physical processes to the descriptions of quantum theory, the explicability of all quantum phenomena in terms of transactions, and a distinction between observable and inferred quantities. He tests the explanatory power of his framework against a number of troubling quantum gedankenexperiments, including the Wheeler delayed-choice experiment, and the Wigner's friend scenario.[107]

Cramer's formulation of the transactional interpretation connects readily to the temporality of determinacy, and is particularly visible in the notion of pseudo-time he uses to explain the transaction between emitter and absorber waves. Referring to this pseudo-time, he claims that this transaction occurs across spacetime intervals as if occurring in a sequence in time outside of the process itself, suggesting a distinction between two orders of time.[108] Not only is Cramer's theory retrocausal to the extent that *advanced* waves propagating from absorber to emitter determine the constituted quantum state; there is a distinction between the time in which advanced and retarded waves are exchanged or transacted—an interaction Cramer labels a "handshake"—accordingly the time of determinacy, and the time used to label or order events. This distinction between the order of events and the order of determinacy, apparent not only in earlier chapters but also in the other quantum-theoretic accounts of retrocausality (such as that of Price), thus recurs in the transactional interpretation. This result gives support to the thesis that acquiring determinacy is, in this interpretation, a process, governed by the dynamics of the transaction. Indeed, Cramer sets out equations which model these transactions for the simplified case of plane offer and confirmation waves in the form $e^{i(kr - \omega t)}$ subject to $t \geq T_1$ or $t \leq T_1$ where T_1 is the instant of emission. Insofar as these equations governing the transaction constitutive of a quantum particle amount to *laws*, insofar as they amount to the theory's *dynamics*, the laws and dynamics which govern the determination process would in this minimal sense depend on time.

One of the principle antinomies of the transactional presentation lies in the following fact. These equations give the impression that these waves propagate in the same time t as applies to the conventional quantum dynamics, and Cramer states this directly. And yet, elsewhere, there are his qualifications which stipulate the parameter to belong to pseudo-time, and that the process itself, despite the occurrence of the parameter t in his most basic model for the transaction, is atemporal.[109] Friederich and Evans, and Lewis, follow this latter view in their studies of the transactional interpretation.[110] Maudlin, in his analysis of Cramer's account, takes this commitment to pseudo-time as a terminal flaw.[111] In his view, in order to be in any way meaningful and in any way serviceable as explanatory devices for quantum processes, sequences of events in pseudo-time must be isomorphic to sequences of events in ordinary time. In such a case, however, he claims that Cramer falls afoul of objections in which *variable* configurations of absorbers, depending sensitively on events preceding them in ordinary time, cannot be regarded as propagating well-defined advanced waves: should the trajectory of the present depend on the future and yet should the future be determined by the present, the theory risks being mired in incoherence.[112] Maudlin generalises his counter-example to any theory where bi-directional influences co-operate,[113] subject to a suitably restricted conception of the term "influence."

Cramer himself appears ambivalent as to the overall ontological status of the program, and summarises his theory in a way which leaves it undecided exactly how seriously his references to these transactions, and the concomitant parameters of pseudo-time, ought to be taken. He refers to it as a model, tool, framework, and conceptual toolkit, and thus a means of visualizing quantum process for pedagogical purposes, with ambitions to dispel counter-intuitive quantum paradoxes.[114] He is less clear as to whether his theory addresses reality. In spite of the prominent resonances of retrocausality within the theory, there is also evidence, then, that Cramer eschews overt ontology and lacks the aspirations of those aiming to define the quantum state outright.

It is therefore debatable as to how seriously transactional quantum theory should be taken as an ontology. However, even were its ontology rejected, this would not amount to a refutation of retrocausality, and still more areas of quantum research and interpretation exist, in addition to those already reviewed, in which retrocausality plays a significant role. Perhaps the most noteworthy case is the two-state vector formalism developed by Aharonov, Bergmann, and Lebowitz. Elitzur and Cohen, in their review of quantum retrocausality, anoint this formalism the most courageous[115] attempt to integrate the concept with the fundamental precepts of quantum theory. This alternative formalism amounts to a time-symmetric presentation of the theory, *pace* the existence of wave-function collapse and branching which might appear to introduce temporal asymmetry into the quantum dynamics. The authors posit in addition to traditional, forwards-evolving quantum state a backwards evolving state which, taken together constitute what they term a two-state vector,[116] which requires the temporal evolution of the quantum state to be reconceived: in the terms of Friederich and Evans, such a device implies that the overall description of the quantum state involves a vector evolving forward in time as well as a vector evolving backwards between boundary conditions.[117] The commitments made by Aharonov, Bergmann, Lebowitz and others in this domain thus link time-symmetry and retrocausality in a way which reproduces the predictions of quantum theory. They achieve as much with relatively minimalistic and restricted ontological commitments in relation to those of other retrocausality theorists such as, for instance, Price.

This section thus attempts to demonstrate that the motivation for abandoning the dogmas of atemporality comes from a number of complimentary directions in the quantum context. The implications of the core of the quantum formalism itself are inseparable from the interpretation one subscribes to; those committed to stochastic collapse processes, as are so-called orthodox or Copenhagen oriented approaches to the theory, as well as dynamical collapse models, have an immediate and obvious relationship to the temporality of determinacy, by virtue of the fact that the randomness of collapse embeds the acquisition of determinacy within the evolution of the quantum state, rather than atemporally. Beyond this core formalism, however, there are other important areas of research which show up the atemporality conjecture as increasingly untenable, particularly in discussions of quantum retrocausality. Huw Price's argument that the plausible (although not indefeasible) assumptions of realism, time-symmetry and discreteness together imply retrocausality provides an example. Whilst Price's conception of retrocausality is for the most part permeated by anthopic notions such as the ability to control or influence events, Leifer and Pusey's more general rigorization of the spirit of Price's argument provides a more precise notion of retrocasaulity, namely the negation of the equality $p(\lambda|\, a, x, y) = p(\lambda|\, a, x)$ and the violation of the following no-retrocausality condition (NRT) for combinations of input and output variables a, b, x, y and physical states λ:

$$p(a,b,\lambda|x,y) = p(b|a,\lambda,x,y)\,p(\lambda|a,x)\,p(a|x).$$
$$(\text{No Retrocausality Thesis NRT})$$

These results match the content of previous chapters, in which a distinction arises between the order of events described by the theory and its laws on the one hand, and the order of their determination on the other. In a different context but with comparable ramifications, Cramer's transactional interpretation of quantum theory envisages exchanges between emitter and absorber particles, with forwards and backwards temporal waves propagating in a pseudo-time sequence. There are, therefore, good reasons to resist the atemporality conjecture in the context of quantum mechanics, in spite of the diversity and fluidity of the prevailing interpretations and the lack of consensus on the formalism's meaning, and there are discernible aspects of the theory which cohere with temporal determinacy.

To conclude, it is worth putting retrocausality aside to explore some of the other, quite distinct sources of insight into time and determinism in quantum theory. As was seen in the introduction to this section, adopting any interpretation of the theory is itself a stance with implications for these, with collapse-oriented approaches introducing stochasticity straight into the quantum dynamics, and others, such as the Everettian approach, claiming the quantum dynamics as paradigms of determinism and the state as a paradigm of determinacy. Aside from the major interpretations of the formalism, though, the innovative field of quantum information theory has also been the site for such analysis. According to one commentator in particular, its implications concern not so much determinacy as a process, or technical distinctions between orders of events and orders of determinacy, but circle back to nomological evolutionism. For Paul Davies, it is not only cosmology which takes laws as well as entities as being subject to change with the universe's emergence and evolution, but quantum information theory also. Davies brings similar questions as to the origin of laws to bear, not only on the tuning of their parameters in a cosmological furnace, but on their very structure, questioning for instance the provenance of two types of electric charge (positive and negative) rather than three or four or more, and of the inverse-squared dependency of Newtonian gravitation and Coulomb's rule rather than any other inverse integer power.[118, 119] Inveighing against what he takes to be the classical, theological and Platonic interpretation of laws according to which these are an unaccountable, base fact[120] he supports Smolin and Wheeler in taking them to be an emergent part of the universe, rather than taking a transcendent position with respect to it.[121] Davies credits Wheeler as a pioneer of this position, and traces back the opposing characterisation of laws—as the prefabricated manifestations of a divine plan[122]—through European theologians and scholars such as Roger Bacon and Thomas Bradwardine. It is quantum information theory which, he claims, enables such a Copernican revolution in the nature of laws.

Davies' argument traverses the following path. First, he outlines several properties put forward by the orthodox conception of laws: they are eternal, immutable, infinitely precise, consecrated in mathematical relationships, and independent of the processes of the world.[123] Second, Davies considers the implications of the claim that these laws, construed as functions, boast infinite precision. Laws, represented by functional relations, are infinitely precise in the sense that the states they describe are representable by quantities whose values are defined to an arbitrary degree of accuracy. A necessary condition for such infinite precision, holds Davies, is there is a physically possible instantiation of every quantity subject to these laws. Yet, according to Davies, there are counter-examples to this latter proposal which are particularly forthcoming in the context of quantum mechanics. Davies considers a series of beam splitters designed either to allow through or to block an incident photon. The ith device is assigned probability p_i of allowing such a photon through. By the normal multiplicative rule, a photon incident on a series of n of these splitters has survival chances given by the product of each probability, namely $p_1p_2\ldots p_n$. For randomly chosen $p_i \in [0, 1]$, Davies claims, there are combinations $p_1p_2\ldots p_n$ which correspond to quantum states for which no physically possible instantiation exists. More specifically, for $n \geq 400$, the complexity of the quantum-mechanical wave-function describing such a state (with an enormity of component states in superposition) would require a system with information content *greater than is available in the entire universe*—an impossibility.

What reasoning leads Davies to this result? Envisaging the universe to include all regions with the ability to interact causally, namely the regions of the universe capable of exchanging light signals, and envisaging the smallest unit of "space" permissible by quantum descriptions (the Planck area[124]), imposes a natural maximum on the universe's informational capacity, given by the relation of the latter to the former. Since Davies takes this well-defined quantum to represent the smallest unit of storage capacity for information instantiating a physical state—a state described by a physical law—the relation between this and the capacity of the causally relatable universe quantifies this maximum. Yet for the n selected in the beam splitter sequence gedankenexperiment, a state of information content of order 10^{122} would be needed to capture the outcome, even for the straightforward case in which $p_i = 0.5$. Davies concludes that the information content of the universe is subject to a theoretical bound, and not merely states of the world that humans perceive,[125] and dismisses the Laplacian image of infinite precision as an absurd anachronism.[126]

What does this conclusion imply as to the status of the laws describing states of this complexity—states with prohibitively large information content? For Davies, these limitations refute the claim that laws are infinitely precise, since a necessary condition for this claim, namely that there is a physically possible instantiation of all quantities they describe, is violated by them. He claims as a result that laws of physics are inapplicable in situations which demand calculations of numbers greater than ~10^{122}.[127] But there is one more step required in order for this line of thought to impinge upon the crucial question as to the temporality of determinacy conferred by these laws: for Davies, the limitation on the precision on laws reflects, not so much epistemic limitations, the clumsiness of experimenters, or the failures of measurement devices, but rather what he terms a minimum of cosmic noise,[128] which is to say an irreducible arbitrariness, stochasticity or perturbation occurring in the laws themselves, *prima facie* not unlike the error terms visible in the time-series models of the previous chapter. This explanation for the failures Davies predicts in our ability to formulate physically meaningful descriptions of the behaviour of complex systems leads to his perspective on the evolutionary character of laws: the laws themselves undergo fluctuations over time, with extreme vicissitudes in the epoch following the big bang, becoming more subtle in later cosmological epochs with increases in the universe's informational capacity. Rather than the predictability or unpredictability of a system reflecting only our knowledge of the system, and nothing of its reality, the system's reality consists in part in what laws can say of it, given its information content; thus, in a Kantian-Copernican reversal, limitations on this content tell us as much about the nature of the laws themselves as they do the states they describe.

Davies' position is controversial, and relies on a number of assumptions. First of all, taking the laws of physics to be laws of information, taking information to encode real, physical states—these are not unproblematic moves. Most importantly for the philosopher interested in the physical sciences' repercussions for the temporality of determinacy, though, is the fact that his approach provides another point of entry to the evolutionary conception of laws, via the lens of quantum information—a conception which leads Davies to offer expansive reflections on the metaphysical relationship between laws and the processes subject to them. The possibility that immutable laws might explain and predict mutable systems Davies dismisses as an atrophied relic of Platonism and theology. Instead, the arrow of dependency and explanation goes in both directions, establishing the kind of reciprocity between laws and processes which befits a self-explanatory universe, whose *explanans* and *explanandum* are contained within one another.[129] In so doing, he brings to bear on the physical sciences a suite of concepts already familiar to students of social-scientific analyses of time-dependency, and echoes, for example, Esposito and Abbott in deploying metaphors of circularity and feedback loops in his characterisation of how this approach leads him to describe the universe.[130] The purported variation over the life of the universe of various critical parameters might encourage one to infer (as have some of the authors mentioned in this chapter) that the functional relations which embed them, relations often at the heart of modern science, are subject to time. From this point of view, the dynamics they encode would in a sense be evolving too, and the movement of celestial bodies over eons would not take place in a theatre governed by static rules imposed from without, but would display patterns which unfold in the drama.

Excursus on Consistency

Conspicuously absent from this chapter's reflections are two of the pinnacles of early Twentieth Century physical science, which have since escaped the confines of academia to penetrate popular culture, with their prediction of miraculous effects such as time dilation and gravitational lensing. First, Einstein's development of the special theory of relativity attempted to use two apparently simple assumptions, a classical principle as to the relativity of motion plus a principle as to the constancy of the vacuum light speed, to resolve foundational inconsistencies between Newtonian mechanics and Maxwellian electromagnetism.[131] Second, his development of the general theory concerns gravitation, and generalises the ethic of the relativity of motion beyond the limited remit of special relativity, which concerns inertial (non-accelerating) references frames. Einstein's simple, fundamental and principle-driven[132] innovations were nonetheless richly generative of novel predictions, since confirmed in spectacular experiments, such as that of Arthur Eddington in 1919, which catapulted the theory into mainstream.

The absence of the special and general theories of relativity from this chapter poses a challenge, not so much to its completeness—which would be unavoidably lacking with or without mention of these theories—but rather to its consistency. Several commentators have drawn attention to the relativity principle employed by Einstein as an objection to any program for time-dependent laws. Lindsay and Margenau in particular submit that the removal of absolute time from laws (understood in their normal way as differential equations) is a requirement of the theory of relativity, and ties this requirement back to the theory's removal of absolute reference frames.[133] Ryckman, in his review of the philosophical foundations of relativity theory, refers to Einstein's principle of general covariance, often taken as one of the foundational principles of general relativity,[134] as a principle of the *form invariance* of the laws of nature, and that co-ordinates (space and time) are arbitrary labels rather than physically meaningful parameters.[135] John Norton, in his extensive review of the topic "General covariance and the foundations of relativity: eight decades of dispute" gives comparable formulations of the same principle. For him, the laws of physics must apply to systems regardless of their state of motion and be amenable to description by equations which are valid for all co-ordinate systems.[136] Julian Barbour in his 2001 "General covariance and best matching" refers to general covariance as a constraint on all viable physical theories.[137]

For this reason, it may be thought that there is an inconsistency between the special or general theories of relativity and the time-dependency of laws. If laws depend on time, it might seem that there must be an absolute simultaneity or non-frame-dependent temporality. How else could one give sense to time-dependency? Yet, this absolute simultaneity is surely incompatible with space and time having the status of co-ordinates, arbitrary labels, just as a non-frame-dependent temporality is surely incompatible with the form invariance of physical laws. This apparent inconsistency clearly needs to be addressed.

Einstein's special and general theories are not the only areas of modern physics which, one might worry, present a conflict with the temporality of determinacy, construed in terms of time-dependent laws.[138] Emmy Noether's theorems in the calculus of variations, published at the same time as general relativity, link energy conservation with the invariance of systems under translations through time. Informally, this move might be thought to inscribe time-translation invariance into the laws describing all systems whose energy is conserved over time. If energy conservation is upheld for all closed systems of interest as one might expect, the one of the four forms of the atemporality conjecture which ascribes this invariance to all laws surely takes on the status of a theorem to which, for systems subject to Noether's theorem, no theory of temporal determinacy could legitimately demur.

The logic of these counter-arguments must be carefully unpacked. First of all, the principle of general covariance has been formulated in a number of different ways which have been contested (as well as its precise relation to general relativity). The title alone of Norton's 1993 paper attests to the best part of a century of dispute over the meaning of the principle, while Brown and Brading open their 2002 paper on general covariance and Noether's theorems with their observation that such debates never seem to recede. Any objection to time-dependency which seeks to use this principle as a catalyst would either have to establish a consensus interpretation of general covariance (a tenuous prospect) or show that a sufficiently wide spectrum of interpretations supported atemporality. Second of all, such an objection would have to point to a logical relationship between this interpretation and the temporality of determinacy, i.e. show that general covariance (as interpreted) entails the atemporality conjectures, or prohibits any one of the various forms of time-dependency elaborated in this work, such as Esposito's distinction between the order in which events are determined and the order in which they occur.

As respects the first of these moves, several different formulations of general covariance must be separated. Clearly the principle has to do in some way with the structure of physical laws, but what constraints does it impose? In his 1916 paper on the foundations of general relativity, "Die Grundlage der allgemeinen Relativitätstheorie," Einstein speaks of the covariance under all possible transformations as a property of laws of nature, which laws hold true independently of co-ordinate systems in which they are expressed. General covariance removes any remaining vestiges of objectivity from space as well as from time.[139] Observations of events concern only coincidences between spacetime points, not the points themselves. The role of physical laws is merely to record and track such coincidences, and so no *a priori* reason exists to favour a particular system of co-ordinates over another; this expresses the principle of general covariance.[140] Einstein summarises this account with the claim that laws should be expressible by equations valid for all co-ordinate systems, and co-variant under any possible substitutions.[141]

Let us summarise the different possible formulations of the principle:

1. There is no *a priori* reason to prefer co-ordinate systems over one another
2. Space and time are not physically objective
3. Laws of nature hold true independent of the co-ordinate systems in which they are expressed

But these three are not the only statements of covariance. As Brown and Brading discuss, Einstein's autobiographical reflections[142] provide evidence that considerations of the simplicity of the mathematical descriptions entered into Einstein's thinking: generally covariant formulations are to be preferred not just for reasons of principle, but because such formulations will typically be the simplest and clearest way in which laws can be expressed.[143] Further, they note, the dazzling variety of the justifications in the 1916 "Foundations" paper for general covariance: Leibnizian arguments, the physical meaning of point-coincidences, and Mach's principle all feature—not to mention Einstein's own presentation of covariance as a natural extension of the relativity principle of special relativity.[144]

The diversity of Einstein's presentation and the ongoing lack of consensus on its ideal formulation should encourage vigilance on the part of any argument espousing the apparent solidity of the principle of general covariance as a refutation of time-dependent laws. What is more, though, for such a refutation to hold good, some interpretation of general covariance, stripped to its bare content, must disqualify time dependency. It is unclear how such an argument can be constructed. Clearly, the independence of laws of nature from the co-ordinate system in which they are expressed and from co-ordinates such as space and time is one of the proposed meanings of general covariance. But only if the term "time" in the phrase "time-dependent laws" or the term "temporality" in "temporality of determinacy" is equated to the time co-ordinate t does this argument function. Is this equation valid?

Subject to context, an affirmative answer to this question *could* be given. This chapter's condensed account of cosmologies in which physical parameters vary by epoch takes this as an indicator that laws can be time-dependent. The meaning of the locution "time" is, here, "cosmological epoch." In Dirac's presentation, this locution was defined in terms of an additional time variable τ, in terms of which the ordinary co-ordinate t itself could be expressed, hence the equation $\tau = \log(t)$. The expansion and aging of the universe unlocks changes in various ratios involving electric charge, atomic masses, Planck lengths and similar. If the variation of parameters such as the gravitational constant G vary with t or τ, laws containing these constants would inherit their time-dependency.

However, there is a sense in which the lessons of general covariance could be used to block this latter logical step. If indeed candidate laws contain time-varying constants, general covariance stipulates that they are ill-formed or, indeed, not laws at all, but merely equations enabling the application of *underlying* laws to defined problems under limited circumstances. Conceived in this aspect, the principle becomes an acid test for whether laws have been cast into their fundamental form and spurious dependencies on frame-dependent co-ordinates extruded. In the case of general relativistic cosmology, Einstein's field equations constitute the fundamental laws. Now, if the constant which determines the strength of attraction between gravitating objects G or the cosmological constant Λ of general relativity varies with the epoch, then must they not imbue the functions in which they appear so prominently with such dependencies as well? On the contrary, precisely the opposite could be argued to be the case. The central dynamical object of general relativistic cosmology is the set of equations known as the Einstein field equations. The common vacuum or matter-free equations take the following form:[145]

$$G_{\mu\nu} \equiv R_{\mu\nu} - \frac{1}{2} R g_{\mu\nu} = 0$$

More generally, these equations often assume the form:

$$R_{\mu\nu} - \frac{1}{2} R g_{\mu\nu} + \Lambda g_{\mu\nu} = \frac{8\pi G}{c^4} T_{\mu\nu}$$

These equations relate constants (such as the gravitational G alleged by Dirac to vary over time) to tensors, mathematical objects (subscripted with the symbols μ, ν in the above) which represent space-time curvature and local distributions of matter, energy and momentum, enabling motion under gravity to be reconceived as free movement of bodies along curved space-time trajectories (geodesics).[146] In general relativity, these equations are fundamental. Approximate solutions to them, derived under certain restrictive assumptions, describe the behaviour of cosmological systems of interest, from approximately Newtonian planetary orbits already well understood by Einstein's time, to novel phenomena discovered in the deep cosmological past, such as the universe's expansion across epochs. Neither the curvature tensor $R_{\mu\nu}$ nor the stress-energy tensor $T_{\mu\nu}$ have any immediately obvious time-dependency. Would these field equations not, then, contain the possibility of *eliminating* time-dependency (such as those of the scale factor $a(t)$ discussed in the first chapter, or those of Dirac's constants discussed in this chapter) from cosmological systems? The fact that parameters change over time does not manifest a deep temporality on the part of the laws containing them, but merely exposes them as a derivative of atemporal laws—in this case, the field equations cited above.

To see this more clearly, take once more the function $a(t)$ which describes the expansion of the universe.

$$\left(\frac{a}{a_{eq}}\right)=\begin{cases}\left(3/2\sqrt{2}\right)^{\frac{2}{3}}\left(H_{eq}t\right)^{\frac{2}{3}} \\ \left(3/\sqrt{2}\right)^{\frac{1}{2}}\left(H_{eq}t\right)^{\frac{1}{2}}\end{cases} \quad (\mathrm{F2})$$

These solutions arise from the following two differential equations:

$$\frac{\dot{a}^2+k}{a^2}=\frac{8\pi G}{3}\rho(t)(\mathrm{F3})$$

$$\frac{2\ddot{a}}{a}+\frac{\dot{a}^2+k}{a^2}=-8\pi Gp(t)(\mathrm{F4})$$

These differential equations, in turn, arise from the Friedmann solutions to Einstein's field equations, namely those which obtain for a universe which is assumed to be (i) *homogeneous* and (ii) *isotropic*[147]—two properties which, whilst enabling useful simplifications, do not hold for our universe. Given a metric which takes the following form, and given a specified matter distribution, the Einstein field equations determine the value of a:

$$ds^2 = dt^2 - a^2(t)\left[\frac{dr^2}{1-kr^2} + r^2\left(d\theta^2 + \sin^2\theta d\varphi^2\right)\right]$$

This development provides some substance for the objection to temporal determinacy. The time-dependency present in the scale factor of the universe arises as a non-static solution to the field equations, applicable subject to assumptions (i) and (ii) or, alternatively, within certain time-scales, i.e. in epochs where the universe has reached significant scales and in which matter distributions become sufficiently thin.[148] This aberrant time-dependency disappears at the level of the field equations, the legitimate dynamical laws of which these causally open systems of co-ordinates are limited solutions. General covariance, taken as the principle that laws of nature are independent of the co-ordinate systems in which they are expressed, can in this sense be viewed as a selection principle, a criterion by which candidate laws can be appraised: should a putative law fail this test, a more fundamental law surely lurks undiscovered. Such might appear to be the implication of, for instance, Friedman's remarks, who takes the principle to eschew physical content in favour of an injunction against formulating theories in a certain way.[149]

Is it fair to deem this a consequence of general covariance? Does this disqualification of time-dependency follow from the principle that the laws of nature be independent of the co-ordinate systems in which they may appear, or the lack of objectivity of time as well as space? Such would appear to be the espousal of Hilbert, who deems physically meaningless any laws without the requisite invariance.[150] But there are reasons to doubt this connection. In particular, it is difficult to reconcile with incongruities between the sense of the term "time" as it appears in Einstein's researches and its usages in the arguments of this work. To begin with the first of these, in the context of relativity theory, time is in effect a label attached to observable events from competing frames of reference. This label is taken from clocks as part of the process of experiment and measurement. In turn, these readings from clocks amount to a counting process: readings of clocks are counts of ticks, which is to say, counts of a process taken to be regular by the observer. In this sense, any process deemed regular, whether planetary orbits or rotations of pointers across watch faces, can function as a waywiser of time. "Time" in the relativistic context therefore reduces to the stipulation of sufficiently high-frequency cyclical processes as being regular enough to function as a reliable tick which can be counted—processes such as atomic decay processes. These define time co-ordinates. In a similar way, the notion of simultaneity, a relation between the time values of distant events, relies upon a procedure by which clocks at the locations of these events are synchronised. This procedure is a convention. Thus, relativity is to be understood as a statement about the symmetries applying to natural laws—being, in the case of special relativity, the symmetries encoded by the Lorentz transformations, which relate inertial frames of reference—then whatever metaphysical commentary is read into its references to time pertains to the relations between clock readings taken in different reference frames. For observers moving in relation to one another at constant speeds, certain mathematical rules discovered by Einstein's Nineteenth Century predecessors relate the spatio-temporal co-ordinates they use to describe their observations; the laws describing these observations must be cast in a form independent of the choice of one particular co-ordinate system over and above another.

This notion of time, then, has a restricted technical sense. Such restrictions establish a gulf in meaning between the usages of the term "time" in the context of relativistic physics over against the temporality of determinacy. In many of the contexts in which this has been discussed in this present work—such as the distinction between the order of events and the order of determinacy at stake in retrocausal quantum models or in the non-physical sciences, or in the stochastic properties of time-series or certain dynamical systems—time signifies something tenuously, if at all, related to the technical usages which crop up in the context of relativity. Accordingly, these examples do not imply the existence of a preferred co-ordinate system in which laws ought to be formulated; they do not require an appeal to objective or absolute time in order to convey their meaning; nor are the laws they describe dependent on the co-ordinate system in which they are expressed. The systems of co-ordinates considered by relativity are foreign to the thought of the theories in question, which have an operational definition, having to do with procedures, such as the stipulation of processes such as ticks as being regular, and the synchronisation of distant clocks.

However, this does not definitively settle this excursus on consistency. In the introduction to this section, relativity was not the only area of Twentieth Century physics which raised suspicions as to the tenability of temporal determinacy. The paragraphs above also mentioned Emmy Noether's research in the foundations of physics, and the relationship between symmetries (such as time translation-invariance) and conservation principles (such as energy conservation) put forward in her 1918 theorems. Does a similar incommensurability hold with this research as was just argued to obtain between the latter and general covariance, or do these results reveal a flaw in the argument of this chapter? Deciding on this question requires a little more detail to be set out. As Brown notes,[151] a view now widespread in the literature (from textbooks to journals) holds that a general and broad relationship holds between symmetries and conservation laws, with an explanatory arrow assumed to exist between the former and the latter. That is, according to this view, the deep mathematical structure of the functional relations which represent the laws of physics ground, account for, undergird the conservation of physical quantities such as energy and momentum, and therefore provide an explanation for the principles familiar to anyone with even loose acquaintance with classical science. In this vein, Bas Van Fraassen, the seminal voice in Twentieth Century philosophy of science, stakes the stark claim that any symmetry will yield a conservation law.[152]

A number of factors, however, militate against premature generalization. To begin with, Noether's 1918 work involves several theorems, not just the first which considers the relationship between symmetries and conservation laws. Her work arises in the context of the calculus of variations, applying to systems which can be described by Lagrangian mechanics. For such systems, the equations and functions which determine the behaviour of the system (and which, under certain circumstances, might be regarded as expressing laws) derive from a mathematical term known as the action which encodes the various forces at work in the given case. This action is subjected to a variational principle known as Hamilton's principle: the action's first-order functional variation, due to arbitrary and infinitesimal transformations of ϕ_k [the field], goes to zero for an arbitrary region of integration, such variations of ϕ_k themselves are zero at the boundary of the integration region.[153] Noether's first theorem considers the implications for its constituent variables of the vanishing of the first-order variation in the action, for a specific smooth infinitesimal transformation applied to these variables. The action is stationary with respect to this infinitesimal transformation applied to the variables. On the basis of this assumption, Noether's reasoning shows that the derivative with respect to time of a narrowly defined function must vanish. Thus, Noether's argument takes the form of a logical conditional, where the antecedent is the assumption that the action is invariant under some variation of the dependent and/or independent variables.[154]

For the purposes of the argument constructed here, there are a number of important lessons to glean from this condensed presentation. Noether's theorem concerns Lagrangian systems for which a well defined action exists, not all equations of motion whatever. Plenty of systems fail to take a Lagrangian formulation, placing them out of scope of the theorem and whatever constraints on the temporality of determinacy are imputed to it; indeed, Brown refers to the theorem as irrelevant in such cases.[155] Even when such systems are amenable to the Lagrangian framework, Noether's result does not entitle one to infer conservation laws in an unqualified fashion: whether or not conservation of energy or momentum can be related to the translation-invariance of space or time depends on the Lagrangian function chosen. Further, in a number of circumstances, the Noether continuity equations can be difficult to interpret as a conservation law. Either they are trivial, mathematically identical with the symmetry, or they lack a meaningful dynamical interpretation as a transition between physical states.

It is therefore no objection to the existence of functional relations which fail to exhibit time translation-invariance that these violate Noether's theorem. The existence of functional relations with the dependencies that have been discussed in this and previous chapters is not in itself incompatible with this theorem. Whether or not conservation principles are violated by time-dependent systems depends sensitively on the dynamics of the system being discussed, and can only be appraised on a case-by-case basis. Indeed, so also does whether or not such violations refute the theory which implies them. Overall, Noether's theorems are complex, and have a restricted technical meaning whose nuances resist being summarily ontologized. Certainly, one must be wary of reading into them the existence of absolute two-way logical arrows between conservation laws and symmetry principles, of the sort that would make such symmetries unassailable by philosophical analysis. What is more, the arguments of this book depend very little on such technical considerations. Rather, they attempt to grasp their relevant metaphysical content—the way in which laws function as explanations given for events, the nature and meaning of determinacy in the context of these laws, and their congruity with properties such as permanence and revisability. It is predominantly on these features of the theories referred to in this chapter that the argument turns, not on constraints on the structure of the action of the theory or the consequences of its extremization.

Broadly speaking, there are two aspects to the argument made in this book: first of all, to demonstrate that a diverse suite of examples, taken from the physical and non-physical sciences as well as the relevant philosophical literature, accommodate a notion of temporal determinacy, and to flesh out the different ways in which they do so; second of all, to demonstrate that they refute a series of conjectures, which are the main ways in which the atemporal conception of laws of nature has appeared in history. Whilst the foregoing paragraphs might or might not succeed in showing the gulf in meaning between the principles of general covariance and Noetherian symmetries and, on the other hand, temporal determinacy in its broader ontological aspect, the question as to the status of the conjectures must also be addressed. Repeating them for alacrity:

> Under the mapping $t \rightarrow t - t_0$ solutions of dynamical equations map to solutions of dynamical equations. (Translation Invariance Thesis)

> For any law f, and any occurrence of any $t_l \in \{t_i\}$, the truth-value of f does not change with variation in t_l. (Truth-Value Constancy Thesis)

> For any law f, f is permanent. (Permanence Thesis)

> For no law f does the time-variable t occur in f where $dg'' + bg' + cg + d = f$ and $g' = \frac{\partial g}{\partial t}$, $g'' = \frac{\partial^2 g}{\partial t^2}$. (Time-Variable Thesis)

The subtleties and qualifications of Noether's research pointed to in these last two paragraphs should also show that her 1918 innovations do not provide a template for a proof of these conjectures to the detriment of the case sketched out herein. If anything, the fact that the differential equations which guide, or represent the guide of, the evolution of systems often derive from a mathematical action principle should also caution authors against taking the absence of t from differential equations as a prerequisite of a well-formed law. The Translation Invariance Thesis does not follow as if a universal tenet, but rather depends on the dynamics of the system in question. The others are ultimately foreign both to the spirit and to the letter of the symmetries and the conservation principles contemplated by these reflections. Thus, just as for the question of the consistency with relativity, the question as to the consistency of temporal determinacy with Noether's work in theoretical physics depends on a number of more subtle technical questions, among them exactly what "link" exists between energy conservation and time-translation-invariance, what form such invariance and conservation principles take (the relation they bear to the atemporality conjecture), and the precise scope of Noether's work in Lagrangian dynamics.

Conclusion

The examples laid out in this chapter illustrate the diverse connotations of temporal determinacy in the physical sciences. Many of them attest to the increasing credibility granted to nomological evolutionism and the variability of laws. Not all relate cosmology; they appear also in other disciplines, such as Davies' theories of quantum information. Whether in Smolin's expansion of time beyond the frontiers of the presently observable universe, or in proposals to experimentally validate fluctuations in basic physical parameters *within* this universe, laws become evolving descriptions—yet, according to their adherents, no less empirically meaningful or falsifiable. Whether or not this dependency is clearly quantifiable, whether or not time variables such as t or τ surface in the corresponding equations, and whatever the status of this or that atemporality conjecture, the evolutionary character of laws inserts time-dependency into the dynamics of the systems in question. This then encompasses the determinacy they impart to physical states.

Considering how important it is to the argument of this section, it is natural to challenge this evolutionism on an obvious question. If parameter variation is what makes physical laws evolve, what of the possibility of deeper laws without such variation? In such a case, the laws thought to contain changing parameters or natural selection mechanisms would resolve into underlying variables without temporal variation. The changes in the parameters and the natural selection mechanisms would themselves be explained away by further time-independent functions. This question was explored briefly in the conclusion of the third chapter: with analyses of complex systems, the temptation arises to think that, recast at the fundamental or microscopic level, such time-dependency may disappear. In the case of the social sciences, this chapter pointed to the fact that there was no precedent for such a reduction, nor any truly fundamental theory in history, and yet no shortage of legitimate analysis of the theories that we have. But does this conclusion still stand up in the physical sciences?

Generally, the ability to resolve diverse laws into simpler, fundamental forms is an important desideratum. The previous sub-section discussed the possibility that Einstein's field equations, as fundamental equations in general relativistic cosmology, could be viewed as the time-independent basis for seemingly time-dependent dynamical laws. Thus, whereas the scale factor $a(t)$ has time-dependency, the field equations in terms of $R_{\mu\nu}$ and $T_{\mu\nu}$ do not:

$$\left(\frac{a}{a_{eq}}\right) = \begin{cases} \left(\frac{3}{2\sqrt{2}}\right)^{\frac{2}{3}} \left(H_{eq}t\right)^{\frac{2}{3}} \\ \left(\frac{3}{\sqrt{2}}\right)^{\frac{1}{2}} \left(H_{eq}t\right)^{\frac{1}{2}} \end{cases} \text{(F2)}$$

$$\frac{\dot{a}^2 + k}{a^2} = \frac{8\pi G}{3}\rho(t)\,\text{(F3)}$$

$$\frac{2\ddot{a}}{a}+\frac{\dot{a}^2+k}{a^2}=-8\pi Gp(t)\,(\mathrm{F4})$$

$$R_{\mu\nu}-\frac{1}{2}Rg_{\mu\nu}+\Lambda g_{\mu\nu}=\frac{8\pi G}{c^4}T_{\mu\nu}$$

General relativistic cosmology, in its seeming reduction of time-dependent expressions to non-time-dependent ones, gives some support to this objection to nomological evolutionism.

Once considered more generally, however, this counter-argument fails. Undermining the conclusions of this chapter requires the *various* models in which laws evolve with systems to be ruled out in favour of static ones which are (in an unelaborated sense) deeper, simpler, explanatorily superior, or more fundamental. In Smolin's model, it is clear that this is not a possibility. The fluctuations of constants at work in the laws of cosmology at the inception of a given universe are random in nature, as he makes plain.[156] They result from minute displacements of the associated points in parameter space from parent universe to child universe, and the selection mechanism which overweights universes with particular matter distributions operates over large numbers of these. Smolin's model and the sense it imparts to temporal determinacy are independent of the particulars of these displacements, which are not subject to any dynamical analysis in his account, nor is there any natural entry point for one, whether or not it were in principle possible. Thus, the claim that any evolving laws could simply be reduced to non-evolving ones, in an attempt to defease the sense in which they support temporal determinacy, whilst not false *a priori*, remains unjustified conjecture in this context.

Evolutionism (with objections to it such as this) is only one of the ways in which the temporality of determinacy makes an appearance in the physical sciences. As lso explored in this chapter, the field of statistical mechanics, long the unwitting victim of philosophers' tussles over issues in the philosophy of time, poses clear challenges to the atemporality conjecture. Various types of ergodic behaviour arise in this field, such as that of mixing or Bernoulli systems, along with the associated property of probabilistic irrelevance. For such systems, the probability of occurrence of events *given* other events is the subject matter of the functions which make up many of the theory's laws. In the appropriate limit, probabilistic irrelevance makes the events described by laws independent of one another, undermining the possibility of atemporal determinative connections between them.

Elsewhere, in the philosophy of quantum mechanics, debates over the temporality of determinacy show up under guises. To begin with, the competing interpretations of the formalism—some of which supplement this formalism with additional rules and variables, others which leave it unperturbed—take different stances on the transition between superposition states and measurement outcomes (if, indeed, they accept that such a transition occurs), such a transition being, for many authors, the point in the evolution of the quantum state at which determinacy is acquired. The temporality of determinacy thus appears as an element of some, though not all, interpretations of the theory. On top of this, though, are the retrocausal aspects of the ontology of the theory, posited by Price, Leifer, Pusey and others. Here, it is claimed that causal or quasi-causal influences propagate from future to past events; more precisely, though, the assumption that earlier states are probabilistically independent of later states are violated for certain quantum experiments. Retrocausality also appears as a feature of specific interpretations, such as Cramer's transactional interpretation with its additional parameter of pseudo-time, and, in Davies' reflections on quantum information.

This chapter has aimed to more fully explore these lines of thought, demonstrating both that good reasons to doubt the validity of the atemporality conjecture can be gleaned from a survey of various fields in the physical sciences, and that detectable marks of temporal determinacy and time-dependent laws can be located there too.

Notes

1. Alborn (1989), p. 1.
2. Whitehead (1978), p. 91.
3. Whitehead (1926), p. 134.
4. Feyerabend (1989), pp. 393–406.

5. For the former see e.g. Faye (2018), or Price (1996), pp. 52–54 for a critical account; for the latter see Wheeler and Zurek (1983), p. 184.
6. Wheeler (1978), p. 3.
7. Wheeler and Zurek (1983), p. 210.
8. Wheeler (1978), p. 11.
9. Wheeler and Zurek (1983), pp. 190, 194.
10. Wheeler and Zurek (1983), p. 210.
11. For an analysis of this common expression see Wallace (2016).
12. Wheeler and Zurek (1983), p. 210.
13. Wheeler (1978), pp. 12–13.
14. Wheeler (1978), p. 11.
15. See Smolin (1992), pp. 173–177. More formally, Smolin defines a point p, akin to a set of parameters, in a parameter space P, and the number N which describes the generation of a given universe, and considers the distribution $\rho(p, N)$, with a view to showing that this function peaks with p maximizing the number of black holes.
16. Smolin (1992), pp. 176, 178, 189. Smolin's instinct is that, our known universe being near to a local maximum of the function which describes the likely number of black holes contained in each, variation of the parameters appearing in fundamental laws should decrease the expected number of black holes.
17. Smolin (1992), pp. 175.
18. Smolin presents this ambition as an answer to the following question as to why the universe is governed by one set of laws rather than another, and what selection process if any led to such a result. Smolin (2013), p. 97. It should be noted that the primary sense in which his approach to cosmological natural selection explains laws is the sense in which it explains the parameters of these laws; Smolin considers variations in parameters rather than variations in other, structural aspects of laws. This could be construed as a significant limitation on the explanatory power of Smolin's model—and has been so construed by e.g. Ladik (2000).
19. Lindsay and Margenau (1963), p. 522.
20. Smolin (2013), p. 97.
21. Lindsay and Margenau (1963), p. 523.
22. Degl'Innocenti et al. (1996), p. 345.
23. Barrow (1996), pp. 1397–1406.
24. Barrow et al. (2002), pp. 201–210.

25. See, among many others, Massa (1993), Godone et al. (1993), Varshalovich and Potekhin (1996), Abdussattar and Vishwakarma (1997), Livio and Stiavelli (1998), Dzuba et al. (1999a), Barabash (2000), Flambaum (2001), Dvali and Zaldarriaga (2002), Landau and Vucetich (2002), Barabash (2003), Langacker (2004). Kozlov et al. (2004), Kafatos et al. (2005), Chiba et al. (2007). Kanekar (2008), Fritzsch (2008), Kajita and Moriwaki (2009) and Mosquera and Civitarese (2011).
26. Dirac (1938), p. 201.
27. Dirac (1938), p. 201.
28. Dirac (1938), p. 201.
29. Dirac (1937), p. 323.
30. Thus, he claims that large numbers must not be viewed as constants, but rather as functions of the epoch…all large numbers of orders 10^{39}, 10^{78}, … which crop up in physical theories are in effect equal to t, t^2… In this model, t is the time of the present epoch in atomic units. See Dirac (1937), p. 323.
31. Kafatos et al. (2005), p. 3142.
32. Dzuba et al. (1999b).
33. s Barrow and Tipler (1986), pp. 20–21, 231–235.
34. Dirac (1938), p. 201.
35. Dirac (1937), p. 323.
36. Barrow and Tipler (1986), pp. 239–242.
37. Callender (2016b).
38. Callender (2016b).
39. Earman (1986), p. 138.
40. Albert (2000), p. 8.
41. More formally, these possibilities are consequences of a characteristic known as time-reversal invariance, familiar from discussions in the previous chapter "Determinacy and Functional Relations." Again, summarily, this characteristic is marked by the property that the mapping $t \rightarrow -t$ translates sets of solutions of the putatively time-reversal invariant dynamical equations to sets of solutions. Heuristically and more loosely, Leifer and Pusey describe how time symmetry means that a video played of a particular process would not allow an observer to tell in which direction it was running. See Leifer and Pusey (2017).
42. Sklar (1993), p. 21.
43. Earman (2006).

44. Barbour (1999), p. 342.
45. Earman (2006).
46. Brown and Myrvold (2008).
47. Frigg et al. (2020).
48. Werndl (2011).
49. Frigg and Werndl (2010), p. 633.
50. Frigg and Werndl (2010), p. 633.
51. Werndl and Frigg (2015), p. 4.
52. See Frigg et al. (2020), Werndl and Frigg (2015), Albert (2000) for a range of contrasting views on the role of ergodicity in Boltzmannian statistical mechanics.
53. Werndl (2009a), p. 214.
54. Frigg et al. (2020).
55. Technically, this definition defines a stronger notion of mixing; a weaker notion, which permits small fluctuations in the concentration over the volume of substances, is also employed. See e.g. Frigg et al. (2020).
56. Werndl (2009a), p. 213.
57. Werndl (2016), p. 3.
58. Werndl (2009a), p. 217.
59. Werndl (2016), p. 1.
60. Spekkens (2005); Harrigan and Spekkens (2010).
61. Hoefer (2016).
62. Russell (1918), p. 199.
63. Earman (2004), p. 8.
64. Interestingly, Boscovich also developed a theoretical presentation of determinism which foreshadowed Laplace's thought experiment involving the demon. See in this respect Kožnjak (2015).
65. See e.g. Hawley (1998), pp. 103–105.
66. More precisely, sets of positive self-adjoint operators whose elements sum to unity represent measurements. See e.g. Wallace (2016).
67. Darby (2010), p. 233.
68. Albert (1992), p. 11.
69. Dirac (1930), pp. 10–14.
70. Barrett (1999), p. 27.
71. It should be noted that the equivalence between eigenstates and determinate values of operators known as the eigenvector-eigenvalue link, whilst extremely prevalent in foundational presentations of quantum mechanics, has been questioned, notably in Wallace (2016).

72. Wilson (2013).
73. Lowe (1994), pp. 110–111.
74. Bub (2000), pp. 598–601.
75. Bell (1987), p. 225, Myrvold (2018).
76. Dirac (1930), p. 36.
77. Bell (1987), p. 215.
78. Wallace (2016).
79. Barrett (1999), p. 54.
80. It is also worth noting that, in spite of its frequent appearance in the commentary, some of the founders of quantum mechanics distanced themselves from the simplistic claim that measurement produces determinacy from superposition, which Wallace refers to as the orthodox view of the orthodoxy of quantum theory. Bohr, for one, acknowledges the immense challenges in coherently separating measurement apparatus from phenomena measured, and questions the relevance of traditional notions of explanation, causality and determinacy altogether. See Bohr (1950), pp. 51–54.
81. Penrose (1996).
82. Ghirardi et al. (1986), p. 480.
83. Bell (1987), pp. 14–21.
84. Pusey et al. (2012).
85. DeWitt (1970).
86. Wallace (2012), p. 13.
87. Price (2012), Price (2008), and Price and Wharton (2013).
88. Price (2012), p. 10.
89. Price (2012), p. 9.
90. Leifer and Pusey (2017), p. 2.
91. Bell (1987), p. 52. Bell gives as an example the potentials of classical electromagnetism, invariant up to a gauge transformation.
92. Price and Wharton (2013), pp. 12–15.
93. Price (1984).
94. Price (1996).
95. Price (2012), p. 10.
96. Russell (1918), p. 199.
97. Leifer and Pusey (2017).
98. Leifer and Pusey (2017), p. 1.
99. Leifer and Pusey (2017), p. 6.
100. Leifer and Pusey (2017), p. 9.

101. Leifer and Pusey (2017), p. 7.
102. Adlam (2018). Adlam's presentation follows that of Bell (1987), pp. 232–248.
103. Kim et al. (1999).
104. Friederich and Evans (2019).
105. Cramer (1987).
106. For a wavefunction, the modulus squared, or the product of the wavefunction with its complex conjugate, corresponding to forward and backwards propagating waves, yields a probability distribution for the position of the particle being described.
107. Cramer (1987).
108. Cramer (2015), p. 7.
109. Lewis (2013).
110. Friederich and Evans (2019).
111. Maudlin (2011), pp. 180–184, 187, 196, 202, 214.
112. Maudlin (2011), p. 184.
113. Maudlin (2011), p. 184.
114. Cramer (1987).
115. Elitzur and Cohen (2011).
116. Aharonov and Vaidman (2007).
117. Friederich and Evans (2019).
118. Davies (2010), p. 68.
119. Aside from his reflections on quantum information, Davies follows the authors cited earlier in embracing the notion of time-varying physical constants, such as the fine-structure constant. See e.g. Davies et al. (2002), p. 602.
120. Davies (2010), p. 68.
121. Davies (2010), p. 83.
122. Davies (2008), p. 7.
123. Davies (2010), p. 70.
124. Davies (2010), p. 77.
125. Davies (2010), p. 82.
126. Davies (2010), p. 74.
127. Davies (2010), p. 84.
128. Davies (2010), p. 86.
129. Davies (2010), pp. 70–73.
130. Davies (2008), p. 251, Davies (2010), p. 87.
131. Other assumptions, such as the isotropy of space, were also crucial to Einstein's argument.

132. Einstein's own thinking on the principle-driven or dynamical characterisation of special relativity changed subtly throughout his life. For a dynamical account see e.g. Brown (2005) and Brown and Read (2019).
133. Smolin (2013), p. 523.
134. Indeed, Einstein at first did so himself, precipitating serious disputes. See e.g. Norton (1993) and Pooley (2017).
135. Ryckman (2018).
136. Norton (1993).
137. Barbour (2001), pp. 199–212.
138. This challenge may indeed be related to the previous one—Brown and Brading draw pedagogical connections between the theorems and what they refer to as coordinate generality, a particular interpretation of general covariance. See Brown and Brading (2002), pp. 59–60.
139. Einstein and Minkowski (1920), p. 97.
140. Einstein and Minkowski (1920), p. 98.
141. Kox et al. (1997), p. 153.
142. Einstein (1970).
143. Brown and Brading (2002), pp. 80–81.
144. Norton (1993), pp. 810–811.
145. Brown (2005), p. 153.
146. Grøn and Hervik (2007), pp. 179–180.
147. Padmanabhan (1993), p. 50.
148. Padmanabhan (1993), p. 50.
149. Friedman (1983), p. 55.
150. Norton (1993), pp. 805–806.
151. Brown (2020).
152. Van Fraassen (1989), p. 258.
153. Brading and Brown (2003), pp. 91–92.
154. Brading and Brown (2003), p. 93, Brown (2020).
155. Brown (2020), p. 4.
156. Smolin (1992), p. 173.

Bibliography

Abdussattar, & Vishwakarma, R. (1997). Some Robertson–Walker Models with Variable G and Λ. *Australian Journal of Physics, 50*, 893–901.

Adlam, E. (2018). Spooky Action at a Temporal Distance. *Entropy, 20*, 41.

Aharonov, Y., & Vaidman, L. (2007). The Two-State Vector Formalism of Quantum Mechanics: An Updated Review. Retrieved December 13, 2020, from https://arxiv.org/pdf/quant-ph/0105101.pdf

Albert, D. (1992). *Quantum Mechanics and Experience*. Harvard University Press.

Albert, D. (2000). *Time and Chance*. Harvard University Press.

Alborn, T. (1989). Peirce's Evolutionary Logic: Continuity, Indeterminacy, and the Natural Order. *Transactions of the Charles S. Peirce Society, 25*(1), 1–28.

Barabash, A. (2000). Possible Evidence for a Time Variation of the Weak-Interaction Constant from Double-Beta-Decay Experiments. *Physics of Atomic Nuclei., 63*(7), 1210–1213.

Barabash, A. (2003). Possible Evidence of Time Variation of Weak Interaction Constant from Double Beta Decay Experiments. *Astrophysics and Space Science, 284*(4), 607–612.

Barbour, J. (1999). *The End of Time: The Next Revolution in Physics*. Oxford University Press.

Barbour, J. (2001). On General Covariance and Best Matching. In C. Callender & N. Huggett (Eds.), *Physics Meets Philosophy at the Planck Scale* (pp. 199–212). Cambridge University Press.

Barrett, J. (1999). *The Quantum Mechanics of Minds and Worlds*. Oxford University Press.

Barrow, J. (1996). Time-varying *G*. *Monthly Notices of the Royal Astronomical Society., 282*(4), 1397–1406.

Barrow, J., Magueijo, J., & Sandvik, H. (2002). A Cosmological Tale of Two Varying Constants. *Physics Letters B, 541*(3–4), 201–210. https://arxiv.org/abs/astro-ph/0204357

Barrow, J., & Tipler, F. (1986). *The Anthropic Cosmological Principle*. Oxford University Press.

Bell, J. (1987). *Speakable and Unspeakable in Quantum Mechanics: Collected Papers on Quantum Philosophy*. Cambridge University Press.

Bohr, N. (1950). On the Notions of Causality and Complementarity. *Science, 111*(2873), 51–54.

Brading, K., & Brown, H. (2003). Symmetries and Noether's theorems. In K. Brading & E. Castellani (Eds.), *Symmetries in Physics: Philosophical Reflections* (pp. 89–109). Cambridge University Press.

Brown, H. (2005). *Physical Relativity: Space-Time Structure from a Dynamical Perspective*. Oxford University Press.

Brown, H. (2020). Do Symmetries "Explain" Conservation Laws? The Modern Converse Noether Theorem vs. Pragmatism. In J. Read, B. Roberts, & N. Teh

(Eds.), *The Physics and Philosophy of Noether's Theorems*. Cambridge University Press. Retrieved December 28, 2020, from https://arxiv.org/pdf/2010.10909.pdf

Brown, H., & Brading, K. (2002). General Covariance from the Perspective of Noether's Theorems. *Diálogos, 79*, 59–86.

Brown, H., & Myrvold, W. (2008). Boltzmann's H-theorem, Its Limitations, and the Birth of (Fully) Statistical Mechanics. Retrieved May 23, 2020, from https://arxiv.org/abs/0809.1304

Brown, H., & Read, J. (2019). The Dynamical Approach to Spacetime Theories. In E. Knox & A. Wilson (Eds.), *The Routledge Companion to Philosophy of Physics*. Routledge.

Bub, J. (2000). Indeterminacy and Entanglement: The Challenge of Quantum Mechanics. *The British Journal for the Philosophy of Science, 51*, 597.

Callender, C. (2016b). Thermodynamic Asymmetry in Time. In E. Zalta (Ed.), *The Stanford Encyclopedia of Philosophy*. Retrieved April 12, 2020, from https://plato.stanford.edu/archives/win2016/entries/time-thermo/

Chiba, T., et al. (2007). Time-Variation of the Proton-Electron Mass Ratio and the Fine Structure Constant with a Runaway Dilation. *Physical Review D, 75*, 10.

Cramer, J. (1987). The Transactional Interpretation of Quantum Mechanics. *Reviews of Modern Physics, 58*, 647–688.

Cramer, J. (2015). The Transactional Interpretation of Quantum Mechanics and Quantum Nonlocality. Retrieved December 8, 2020, from https://arxiv.org/pdf/1503.00039.pdf

Darby, G. (2010). Quantum Mechanics and Metaphysical Indeterminacy. *Australasian Journal of Philosophy, 88*(2), 227–245.

Davies, P. (2008). *The Goldilocks Enigma: Why Is the Universe Just Right for Life?* Houghton Mifflin.

Davies, P. (2010). Universe from Bit. In P. Davies & N. Gregersen (Eds.), *Information and the Nature of Reality* (pp. 65–91). Cambridge University Press.

Davies, P., Davis, T., & Lineweaver, C. (2002). Cosmology—Black Holes Constrain Varying Constants. *Nature, 418*, 602–603.

Degl'Innocenti, S., et al. (1996). Time-Variation of Newton's Constant and the Age of Globular Clusters. *Astronomy and Astrophysics, 312*, 345–352.

DeWitt, B. (1970). Quantum Mechanics and Reality. *Physics Today, 23*(9), 30–35.

Dirac, P. (1930). *Principles of Quantum Mechanics*. Oxford University Press.

Dirac, P. (1937). The Cosmological Constants. *Nature, 139*, 323.

Dirac, P. (1938). A New Basis for Cosmology. *Proceedings of the Royal Society, 165*(921), 199–208.

Dvali, G., & Zaldarriaga, M. (2002). Changing α with Time: Implications for Fifth-Force Type Experiments and Quintessence. *Physical Review Letters, 88*, 091303.

Dzuba, V., Flambaum, V., & Webb, J. (1999a). Calculations of the Relativistic Effects in Many-Electron Atoms and Space-Time Variation of Fundamental Constants. *Physical Review A, 59*(1), 230–237.

Dzuba, V., Flambaum, V., & Webb, J. (1999b). Space-Time Variation of Physical Constants and Relativistic Corrections in Atoms. *Physical Review Letters, 82*(5), 884.

Earman, J. (1986). *A Primer on Determinism*. D. Reidel.

Earman, J. (2004). Determinism: What We Have Learned and What We Still Don't Know. Retrieved April 10, 2020, from https://www.ucl.ac.uk/~uctytho/Determinism.pdf

Earman, J. (2006). The "Past Hypothesis": Not Even False. *Studies in History and Philosophy of Modern Physics., 27*(3), 399–430.

Einstein, A. (1970). Autobiographical Notes. In A. Schlipp (Ed.), *Einstein: Philosopher-Scientist*. Open Court.

Einstein, A., & Minkowski, H. (1920). *The Principle of Relativity* (M. Saha & S. Bose, Trans.). University of Calcutta.

Elitzur, A., & Cohen, E. (2011). The Retrocausal Nature of Quantum Measurement Revealed by Partial and Weak Measurements. *AIP Conference Proceedings, 120*, 1408.

Faye, J. (2018). Backward Causation. In E. Zalta (Ed.), *The Stanford Encyclopedia of Philosophy*. Retrieved October 4, 2020, from https://plato.stanford.edu/archives/sum2018/entries/causation-backwards/

Feyerabend, P. (1989). Realism and the Historicity of Knowledge. *Journal of Philosophy, 86*, 393–406.

Flambaum, V. (2001). Does the Fine Structure Constant Vary with Time and Distance? In *Atomic Physics 17: International Conference on Atomic Physics* (Vol. 551, pp. 86–99). American Institute of Physics.

Friederich, S., & Evans, P. (2019). Retrocausality in Quantum Mechanics. In E. Zalta (Ed.), *The Stanford Encyclopedia of Philosophy*. Retrieved December 6, 2020, from https://plato.stanford.edu/entries/qm-retrocausality

Friedman, M. (1983). *Foundations of Space-Time Theories: Relativistic Physics and Philosophy of Science*. Princeton University Press.

Frigg, R., Berkovitz, J., & Kronz, F. (2020). The Ergodic Hierarchy. In E. Zalta (Ed.), *The Stanford Encyclopedia of Philosophy*. Retrieved November 1, 2020, from https://plato.stanford.edu/archives/fall2020/entries/ergodic-hierarchy/

Frigg, R., & Werndl, C. (2010). Explaining Thermodynamic-Like Behaviour in Terms of Epsilon-Ergodicity. *Philosophy of Science, 78*, 628–652.
Fritzsch, H. (2008). The Fundamental Constants in Physics and Their Time Dependence. *Progress in Particle and Nuclear Physics., 621*(2), 329–342.
Ghirardi, G., Rimini, A., & Weber, T. (1986). Unified Dynamics for Microscopic and Macroscopic Systems. *Physical Review D, 34*(2), 470–491.
Godone, A., et al. (1993). New Experimental Limits to the Time Variations of $g_p(m_e/m_p)$ and α. *Physical Review Letters, 71*(15), 2364–2366.
Grøn, Ø., & Hervik, S. (2007). *Einstein's General Theory of Relativity: With Modern Applications in Cosmology*. Springer Science & Business Media.
Harrigan, N., & Spekkens, R. (2010). Einstein, Incompleteness, and the Epistemic View of Quantum States. *Foundations of Physics, 40*, 125–157.
Hawley, K. (1998). Indeterminism and Indeterminacy. *Analysis, 58*(2), 101–106.
Hoefer, C. (2016). Causal Determinism. In E. Zalta (Ed.), *The Stanford Encyclopedia of Philosophy*. Retrieved May 2, 2020, from https://plato.stanford.edu/archives/spr2016/entries/determinism-causal/
Kafatos, M., Roy, S., & Roy, M. (2005). Variation of Physical Constants, Redshift and the Arrow of Time. *Acta Physica Polonica, 36*(10), 3139–3161.
Kajita, M., & Moriwaki, Y. (2009). Proposed Detection of Variation in m_p/m_e Using a Vibrational Transitional Frequency of a CaH^+ Ion. *Journal of Physics B, 42*, 6.
Kanekar, N. (2008). Do Fundamental Constants Change With Time? *Modern Physics Letters A, 23*(32), 2711–2725.
Kim, Y., et al. (1999). A Delayed Choice Quantum Eraser. *Physical Review Letters, 83*, 1–5.
Kox, A., Klein, M., & Schulmann, R. (1997). *The Collected Papers of Albert Einstein, Volume 6, The Berlin Years: Writings, 1914–1917*. Princeton University Press.
Kozlov, M., et al. (2004). Space-Time Variation of the Fine-Structure Constant and Evolution of Isotope Abundances. *Physical Review, A, 70*, 6.
Kožnjak, B. (2015). Who Let the Demon Out? Laplace and Boscovich on Determinism. *Studies in the History and Philosophy of Science, 51*, 42–52.
Ladik, J. (2000). Did the Universe Evolve by Chance? Remarks to Lee Smolin's Book "The Life of the Cosmos". *Journal of Molecular Structure: Theochem, 500*(1–3), 1–3.
Landau, S., & Vucetich, H. (2002). Testing Theories That Predict Time Variation of Fundamental Constants. *Astrophysics Journal, 570*(2), 463–469.

Langacker, P. (2004). Time Variation of Fundamental Constants as a Probe of New Physics. *International Journal of Modern Physics, 19*, 157–165.
Leifer, M., & Pusey, M. (2017). Is a Time Symmetric Interpretation of Quantum Theory Possible Without Retrocausality? *Proceedings of the Royal Society A, 473*, 2202. Retrieved October 31, 2020, from https://arxiv.org/pdf/1607.07871.pdf
Lewis, P. (2013). Retrocausal Quantum Mechanics: Maudlin's Challenge Revisited. *Studies in History and Philosophy of Science Part B: Studies in History and Philosophy of Modern Physics, 44*(4), 442–449.
Lindsay, R., & Margenau, H. (1963). *Foundations of Physics*. Dover Publications.
Livio, M., & Stiavelli, M. (1998). Does the Fine-Structure Constant Really Vary in Time? *The Astrophysics Journal, 507*, L13.
Lowe, E. (1994). Vague Identity and Quantum Indeterminacy. *Analysis, 54*(2), 110–114.
Massa, C. (1993). A New Limit on Time Dependence of the Gravitation al Constant from Gravity Modified Quantum Electrodynamics. *Astrophysics and Space Science, 209*, 307–308.
Maudlin, T. (2011). *Quantum Non-Locality and Relativity: Metaphysical Intimations of Modern Physics* (3rd ed.). John Wiley & Sons.
Mosquera, M., & Civitarese, O. (2011). Time Variation of the Fine Structure Constant and of the Higgs Vacuum Expectation Value on Cosmological Time Scales. *Astronomy and Astrophysics, 526*, 1–4.
Myrvold, W. (2018). Philosophical Issues in Quantum Theory. *The Stanford Encyclopedia of Philosophy*. Retrieved November 20, 2020, from https://plato.stanford.edu/archives/fall2018/entries/qt-issues/
Norton, J. (1993). General Covariance and the Foundations of General Relativity: Eight Decades of Dispute. *Reports on Progress in Physics, 56*, 791–858.
Padmanabhan, T. (1993). *Structure Formation in the Universe*. Cambridge University Press.
Penrose, R. (1996). On Gravity's Role in Quantum State Reduction. *General Relativity and Gravitation, 28*, 581–600.
Pooley, O. (2017). Background Independence, Diffeomorphism Invariance, and the Meaning of Coordinates. In D. Lehmkuhl, G. Schiemann, & E. Scholz (Eds.), *Towards a Theory of Spacetime Theories* (pp. 105–144). Springer.
Price, H. (1984). The Philosophy and Physics of Affecting the Past. *Synthese, 61*(3), 299–323.

Price, H. (1996). *Time's Arrow and Archimedes' Point: New Directions for the Physics of Time*. Oxford University Press.
Price, H. (2008). Toy Models for Retrocausality. *Studies in History and Philosophy of Modern Physics, 39*(4), 752–761. Retrieved January 12, 2020, from https://arxiv.org/pdf/0802.3230.pdf
Price, H. (2012). Does Time-Symmetry Imply Retrocausality? How the Quantum World Says "Maybe". *Studies in the History and Philosophy of Modern Physics, 43*, 75–83.
Price, H., & Wharton, K. (2013). Dispelling the Quantum Spooks—a Clue that Einstein Missed? Retrieved November 11, 2020, from https://arxiv.org/pdf/1307.7744.pdf
Pusey, M., Barrett, J., & Rudolph, T. (2012). On the Reality of the Quantum State. *Nature Physics, 8*, 475–478. Retrieved November 11, 2020, from https://arxiv.org/pdf/1111.3328.pdf
Russell, B. (1918). *Mysticism and Logic and Other Essays*. Longmans, Green and Co.
Ryckman, T. (2018). Early Philosophical Interpretations of General Relativity. In E. Zalta (Ed.), *The Stanford Encyclopedia of Philosophy*. Retrieved December 23, 2020, from https://plato.stanford.edu/entries/genrel-early/
Sklar, L. (1993). *Physics and Chance*. Cambridge University Press.
Smolin, L. (1992). Did The Universe Evolve? *Classical and Quantum Gravity, 9*, 173–191.
Smolin, L. (2013). *Time Reborn*. Penguin.
Spekkens, R. (2005). Contextuality for Preparations, Transformations, and Unsharp Measurements. *Physical Review A, 71*, 052108.
Van Fraassen, B. (1989). *Laws and Symmetry*. Clarendon Press.
Varshalovich, D., & Potekhin, A. (1996). Have the Masses of Molecules Changed during the Lifetime of the Universe? *Astronomy Letters, 5*(1), 1–5.
Wallace, D. (2012). *The Emergent Multiverse: Quantum Theory According to the Everett Interpretation*. Oxford University Press.
Wallace, D. (2016). *What Is Orthodox Quantum Mechanics?* Retrieved October 9, 2020, from https://arxiv.org/pdf/1604.05973.pdf
Werndl, C. (2009a). What Are the New Implications of Chaos for Unpredictability? *The British Journal for the Philosophy of Science, 60*(1), 195–220.
Werndl, C. (2011). Observational Equivalence of Deterministic and Indeterministic Descriptions and the Role of Different Observations. In *EPSA Philosophy of Science: Amsterdam* (pp. 427–439). Springer.

Werndl, C. (2016). Determinism. In M. Griffith, T. Neil, & K. Timpe (Eds.), *Routledge Companion to Free Will.* Routledge. Retrieved May 10, 2020, from http://philsci-archive.pitt.edu/12166/1/Determinism IndeterminismWordPittsburghArchiveWithF.pdf
Werndl, C., & Frigg, R. (2015). Reconceptualising Equilibrium in Boltzmannian Statistical Mechanics and Characterising its Existence. *Studies in History and Philosophy of Modern Physics, 49*, 19–31. Retrieved November 7, 2020, from https://arxiv.org/pdf/1510.02260.pdf
Wheeler, J. (1978). *Frontiers of Time*. University of Texas.
Wheeler, J., & Zurek, A. (1983). *Quantum Theory and Measurement*. Princeton University Press.
Whitehead, A. (1926). *Science and the Modern World: Lowell Lectures*. Cambridge University Press.
Whitehead, A. (1978). *Process and Reality: An Essay in Cosmology*. The Free Press.
Wilson, J. (2013). A Determinable-Based Account of Metaphysical Indeterminacy. *Inquiry, 54*(4), 359.

5

The Temporality of Determinacy III: Kant and Contemporary Philosophy

Introduction

The work of the previous chapters may rightly be regarded as less an original philosophical undertaking, more a work of aggregation: assimilating supporting evidence, stretching across the span of the physical and social sciences, for the temporality of determinacy, and taking these examples as opportunities to test the veracity of metaphysical claims about the temporality of laws. From the evolution of physical parameters at the cosmological scale to recent decades' innovation in cryptography, examples abound of systems which, as described by the best available theories, no longer accommodate time-independent descriptions. These chapters, then, aim not so much to subject temporal determinacy to analysis a priori, but to take as a guiding clue to the question at hand both established and pioneering fields of research in other disciplines. This strategy, however much it hopes to remain grounded and mitigate the potential for unmoored speculation, has an obvious lacuna. What are the broader philosophical ramifications of the findings of this search, aside from the refutation of the atemporality theses defined, in rather a narrow way, in the first chapter?

Certainly, such is not the only omission of the project developed up to here. The entwinement of the atemporality theses with the metaphysics

C. Husbands, *The Temporality of Determinacy*,
https://doi.org/10.1007/978-3-030-86530-6_5

of time and determinism tempts one to pose much broader questions in response: which implications does the denial of these theses harbour for determinism generally? It might, after all, be thought that this denial undermines altogether the authority of laws, and plunges us into a generalized avowal of indeterminism or indeterminacy, in which humanity's understanding of nature is swept away helplessly by a torrent of rhapsodically fluctuating laws, while predictions and models anachronize over time.

It is the working hypothesis of this chapter that the philosophy of Kant provides a suitable opportunity for the exploration of these questions. Specifically, Kant's philosophy amounts to a convergence point of two ostensibly clashing lines of thought: (i) the temporality of determinacy and (ii) determinism, and the uniformity and consistency of nature bound up with it. Thus, his thought provides the conditions of possibility for the reconciliation of determinism (in a sense to be qualified) with the abandonment of the atemporality conjecture, and the adoption of time-dependent laws of nature.

The reader might well ask why Kant of all thinkers would be considered the best representative of these two lines of thinking. Staving off the threat that the Pyrrhonian tendencies of the metaphysics of the time would unleash chaos was of fundamental importance to Kant. Questions in his work as to the uniformity and consistency of nature and the intelligibility of the system of the world are frequently presented as driving forces for the birth of Critical philosophy. As Teufel suggests in his essay "Kant's Transcendental Principle of Purposiveness and the "Maxim of the Lawfulness of Empirical Laws"" Kant's principle of purposiveness appears as a transcendental rule unveiled in order to stave off the risk of experiential chaos, inconstancy and instability.[1] In many places in the *Critique of Pure Reason*, Kant appeals to the need for defence against such a risk as a justification for transcendental propositions. In A100, Kant discusses the consistency of the properties of substances as a necessary condition for the proper functioning of the empirical imagination (specifically, its associative powers): If substances were to change their basic characteristics such as weight or colour, association of experiences by virtue of the imagination would not be possible.[2] In a number of places in this *Critique*, Kant refers to the counter-thesis to transcendental idealism as the reduction of nature to a rhapsody of perceptions: reason requires that cognitions must be systematic and organised rather than rhapsodic and capricious.[3] Experience requires a unity of appearances without which ensues a rhapsodic array of perceptions.[4] Later in the text, it becomes clear that such is also, for Kant, a consequence of reason's legislative prescriptions.[5]

Other well excavated aspects of the *Critique* show that the constancy of the behaviour of objects of scientific enquiry was a central concern for Kant. As will be demonstrated in more detail throughout this chapter, the Schematism of the Understanding and the Analogies of Experience, all display elements of Kant's belief in the subjection of the manifold of intuition[6] to *rules* he takes as a mark of ordered and non-rhapsodic experience. This principle he relates in turn to the doctrine of the unity of apperception, that is, the identity of the self in respect of all possible representations,[7] claiming that experience is grounded a priori by unity-imposing rules in the synthesis of appearances.[8] At the same time, Kant considers the relation of these synthetic facilities which serve as grounds for objectivity to empirical laws, a recurring concern in his writings, from the first and the final *Critiques*, to the unpublished notes, to essays such as the *New Elucidation* and to other standalone monographs. Aside from the conclusions of the Analogies of Experience, such as the law of connection of cause and effect which scholars such as Melnick, Friedman and Guyer have taken to entail the existence of individual causal laws,[9] Kant's opus frequently addresses the interplay between transcendental-philosophical principles and laws which form the grounds of scientific theory (and thus a metaphysics of nature).[10] In the *Metaphysical Foundations of Natural Science*, for instance, Newton's Laws take on the status of conditions of possibility of experience, a status ascribed also to the categories—laws which Beck refers to as natural laws of cosmology, which intermediate between universal, necessary and fundamental laws of nature (such as the connection of cause and effect) and particular empirical laws.[11]

The uniformity and consistency of nature thus constitutes in Kant both a motif and a theme. They crop up in his many references to laws and rules which confer order on experience. And just as the question of this book as to the time-dependency or atemporality of laws of nature is entwined with the way in which the determinacy of nature becomes a process, for Kant, these two characteristics attributed to nature connect to an analysis of time-determination—the ascription of temporal properties to objects or, so to speak, objects' locations in time.[12] Kant speaks of "conditions of complete and necessary determination of time for all existence"[13] as a necessary condition for empirical time-determination locating this determination in the unity of apperception. "All empirical time-determinations," he claims, must be subject to "rules of universal time-determination."[14] The schemata which mediate the application of categories to appearances, a relationship referred to by many early analysts as a centrepiece of the first *Critique*,[15] Kant considers to be transcendental and pure time-determinations.[16] Watkins takes what he describes as a problem of time-determination as one of the driving forces responsible for the Analogies of Experience.[17] In particular, he claims that Kant's very purpose in the Second Analogy is to specify the grounds for objects' temporal determination, a development of his pre-Critical period thought.[18] Kant thus more than perhaps any other thinker of his epoch problematized intensely the temporal status of experience, inferring from the mind's ability to cognize such experience the existence of retentive and imaginative synthetic facilities belonging to the thinking subject. Such a perspective was seized upon by his successors in and around the tradition of the philosophy of process, such as Hegel and Heidegger, Cassirer and Bergson.

Kant's thought in this sense provides a credible benchmark against which the character of the temporality of determinacy can be assessed, by virtue of the sheer stature and great reach of his interventions in the metaphysics of causality and time. This is, however, only a basic indication of the utility of the Kantian literature for the temporality of determinacy. As this chapter will show, Kant's project offers broader resources to address the problems which are generated by the time-dependency of *laws* specifically. These problems vitalized Kant throughout his writings, from the *Critiques* to his unpublished notes. Thus, on the one hand, the concerns raised in the context of philosophical discussions of the temporality of determinacy contemplated its logical relation to the impermanence, changeability, and the usurpation of lawlike behaviour. On the other, the motivating kernel of Kant's project was ensuring the integrity of laws, in the form of their universality and necessity, given the challenges levelled at reason by empiricist skepsis. Thus, he describes his guiding question at the beginning of the *Prolegomena*, of the provenance of necessity in experience, which is not a priori: how is it that the subject is able to go beyond concepts and connect them with others which are not contained within them, in such a way that the latter necessarily belongs to the former?[19] With his response to this question, Kant allows us to trace the emergence of universal and necessary propositions in the metaphysics of experience, and therefore in the foundations of science, to its categorical roots in the understanding. The constitutive function of these categories involves in an essential way the temporal determination of appearances, their transformation from mere intuitions into elements of objective experience with temporal characteristics.

It is for these reasons that this chapter takes Kant's philosophy as a useful lens for studying the broader repercussions of the temporality of determinacy and the attack on the atemporality conjectures built up throughout this work. It considers the three primary areas of the *Critique of Pure Reason* in which his concern with the status of laws of nature and the time-determination of appearances are most clearly discernible:

1. The Schematism of the categories,
2. The argument for the doctrine of apperception
3. The Analogies of Experience.

Subsequently, it addresses other texts which expand upon the themes of time and determination. It considers those aspects of Kant's work which appear *prima facie* to support an atemporal conception of the rules governing appearances, such as the first Analogy, in which the permanence of the underlying ground for the determination of manifold and substrate of change in appearances is asserted explicitly.[20] It argues these conclusions to be oversimplifications of Kant's view and, in contrast, stakes two claims. First of all, viable interpretations of Kant which emphasise his view that the processual *activity* of the mind is responsible for the crucial schematism of the categories—interpretations such as those which Eva Schaper labels constructionalist interpretations—provide room for temporal determinacy to be integrated within a Kantian worldview. The constructive activity of the imagination displays what Kant terms a kind of free lawfulness,[21] viz., conformity to rules in conjunction with spontaneity. Thus, synthesis *exemplifies* rules. Second of all, complementary material from Kant's later works, such as the *Critique of Judgement*, furnishes a dynamic conception of natural laws in which the necessity and universality of laws is linked to the lawful or systematic unity of nature, underpinned by a principle of purposiveness.[22] According to this account, staving off the threat of chaos requires not only taking a certain position with respect to the status of empirical laws but also the adoption of a practical principle, according to which nature exhibits systemic unity. Absent this assumption cognitions would dissolve into contradiction, and no scientific theory could be constituted.[23] The commentaries of Thomas Teufel, Martin Heidegger and Robert Pippin are put forward in support of this interpretation. The chapter concludes that Kant's philosophy opens up a path to reconciliation between the bifurcating aspects of temporal determinacy on the one hand and the desirable characteristics of natural laws, namely consistency and stability, on the other, avoiding the descent of temporal determinacy into unreason and absurdity. Thus, the intuitive principles of consistency and uniformity do not entail the time-independence of the laws being described. The seemingly counter-intuitive notions of temporal determinacy and time-dependent physical laws are, in fact, integrable with established philosophical canon.

Time and Determinacy in the Analogies of Experience

Kant's works are replete with notions of law and lawfulness, from the moral laws explored in the *Critique of Practical Reason* to the laws of nature explored in his pure metaphysics, to the mathematical and logical laws of the pre-Critical works.[24] Controversy exists over the degree of unity or disunity exhibited by these notions, and over the exact scope of Kant's interest in laws. Initially, Kant's own distinction between the moral and metaphysical meanings of laws seems oppositional: in his famously elegant finale to the *Critique of Practical Reason,* he describes the origins of the laws of nature in the contingent external world over against the source of morality in the self which stands in contrast to universal, necessary connections.[25] But scholars need not infer from distinctions such as this that Kant operates with a polysemic conception of law: Watkins, in his *Kant on Laws* and "Kant on the Unity and Diversity of Laws," argues for a univocal one, aiming to show that a unified conception of Kant's notion of laws is possible in spite of the diverse contexts in which they are deployed.[26]

Aside from this question of unification in Kant's theory of laws is one of scope. A supermajority of leading modern Kant scholars have opined on the question as to whether Kant's analysis of laws has anything meaningful to say of empirical laws, or whether it confines itself for the most part to their transcendental undergirding—to metaphysical principles devoid of empirical content—from Buchdahl to Van Cleve, Allison to Guyer. On the one hand, Kant's focus on laws in his Critical writings involves little empirical content: the Second Analogy, for instance, refers to *the* law of cause-effect relations[27] rather than relations between specific cause-effect tokens. On the other, the *Metaphysical Foundations of Natural Science* is a text which seems to imply a correspondence between the transcendental principles of the Analogies and Newtonian laws of mechanics.[28] James Messina distinguishes between broadly three points of view on this: (i) a derivation account, attributable to Michael Friedman, according to which a priori laws of nature ground empirical laws by application to sensible content; (ii) a best system approach, attributable to Philip Kitcher, according to which belief in specific laws is not so much entailed by metaphysical axiom, but pragmatically justified by the need to form an overall theory; (iii) a third way according to which laws supervene on the nature of the systems they describe.[29] As Messina summarises in his review "Kant's Necessitation Account of Laws and the Nature of Natures," much of the debate over Kant's account attempts to relieve the tension between the following propositions: first, empirical laws are not known a priori; second, empirical laws hold in some sense of necessity; third, experience cannot confer such necessity; fourth, some empirical laws are known.[30] It is not obvious that these propositions cohere, yet they all seem to be commitments of Kant's.

As far as these questions are concerned, Kant's Analogies of Experience are perhaps the central site of contention. They consider the synthetic a priori propositions or principles of understanding which result from the application of the categories to representations in time.[31] Previous chapters of the *Critique of Pure Reason* focused on identifying and deriving these categories, which allow objects to be thought, and demonstrate their applicability to appearances. The Analogies proceed to consider a key question left unresolved by these discussions, such as the Transcendental Deduction, namely the grounds of objective cognition of temporal order, and therefore temporal experience.[32] Thus Kant's statement of purpose mentions the need to assess the conditions for a "rule according to which a unity of experience may arise from perception" and for "rules of the determination of their relation to one another in one time."[33] These definitions make the Analogies natural targets for an assessment of Kant's view of the temporality of laws: the conditions for the temporal determination of objects of experience seem on the face of it to constitute a Kantian proxy for the temporality of functional relations.

Kant's First Analogy might seem to pose a problem for any attempt to locate in Kant's work the germination of temporal determinacy, with its semblances of the terminology of the atemporality conjectures. Consider, then, this argument, and its relation to the thesis of this book. Crudely, Kant's argument is that the determination of appearances temporal relations to one another requires a permanent substance which persists throughout changes in them:

> The permanent in relation to which alone all time-relations of appearances can be determined, is substance in the appearance, that is, the real in appearance, and as the substrate of all change remains ever the same.[34]

Objective relations of succession, simultaneity and duration which obtain between appearances require a backdrop or reference frame in which they are determinable.[35] Thus, Kant concludes, there is an unchanging, remaining ground through which all changes in appearances must be thought. Whilst it is contested actively by scholars quite what Kant intends by the substance he takes to provide this ground—whether material substance, of the sort which might be the subject of a conservation law, or a pure category without meaningful connotations in physical science—the fact that its role involves determining appearances' temporal relations is quite clear. For Kant, this entails the work of an

> underlying ground which exists *at all times*, that is, something *abiding* and *permanent*, of which all change and coexistence are only so many ways (modes of time) in which the permanent exists.[36]

The Parmenidean language—casting as it does the substratum of change as abiding, permanent, and omnipresent—cannot fail to stir up vestiges of the permanence thesis. If changes in nature such as are observed in human experience are effected by laws, and such are Kant's grounds which abide permanently, then it would appear that transcendental idealism not only aligns with atemporal determinacy, but incorporates it into its distinctive precepts, among them the First Analogy.

A cursory survey of the Second Analogy may compound these suspicions. Perhaps the most obdurately debated of the Analogies, Kant provides distinct expressions of the principle in the A and B editions of the first *Critique*:

> Everything that happens, that is, begins to be, presupposes something upon which it follows according to a rule.
>
> All alterations take place in conformity with the law of the connection of cause and effect.[37]

A question raised frequently by commentators on the Analogy is one of scope. Does the Analogy harbour ramifications for the causal relations of actual science, and therefore functions, such as Coulomb's Law, Boltzmann's transport equation, or the predictive statistical rules of the social sciences? Allison takes a softer perspective on Kant's argument, claiming that Kant aims to establish only that *some* cause must exist for any event, proposing no specifics as to what this cause must look like, and therefore saying little or nothing as to the status of particular lawlike relations. Friedman's approach, by contrast, holds that such particular causal relations between events of certain types and events of other types, follow from the existence of the Second Analogy[38]—even if the *Critiques* themselves fail to enumerate many of these. For the purposes of the question of temporal determinacy, however, the question as to whether or not specific or particular laws are within the scope of Kant's analysis is only partially relevant. What is more important is that these laws determine the temporal relations of appearances. As a result, the characteristics Kant attributes to them impinge upon his view of temporal determinacy. In order to take this further, there are several such characteristics which must be distinguished.

1. First of all, Kant repeatedly describes the determination effected by causal relations of the Second Analogy as being effected by grounds, and such grounds as *unchanging* grounds. Just as the First Analogy prescribes the need for a permanent in relation to which time-relations can be established, Watkins, for example, asserts unchanging grounds as an important element of Kant's account of the causal relation described in the Second Analogy and imputes to Kant an ontological framework which entails directly that laws cannot change.[39]
2. Second of all, Kant asserts both the universality and the necessity of such a relation. In loose notes from the *Duisburg Nachlaß* written around 1775, Kant refers repeatedly to the universality of whatever law determines a representation to succeed another, and considers the necessary nature of the relations involved as something which follows from universality.[40] Later in these notes Kant equates the Analogies with the requirement for a universal condition which determines appearances temporally.[41] In the *Critique of Judgement*, Kant alludes to the subject matter of the Second Analogy, as a universal law.[42] Indeed, as he never tires of impressing in the *Prolegomena*, the motivation for Kant's early Critical project, namely securing the possibility of synthetic a priori judgements, arises from the validity, unaccountable for until then, of *universal* and *necessary* laws of nature, among them the maxim that all events is determined previously, and by laws which are constant.[43] It is thus in no way surprising that the apotheosis of Kant's focus on causal relationships in the *Critique of Pure Reason* attributes universality to them. This view is supported by the scholarship: for instance, Watkins' unitary account of Kant's notion of laws posits necessity as one of its two defining characteristics.[44]

3. Third, Kant connects these two principles to the generalizability or uniformity of the laws: only by perceiving events which follow one another regularly are rules discovered by virtue of which these events are *always* so connected.[45] In the *Critique of Pure Reason*, Kant claims that rules need a uniformity (*Gleichförmigkeit*) of effects, and that this is the foundation of the concept of cause.[46] In this vein, Beck, for example, singles out the Second Analogy as a possible foundation for the generalized validity of natural laws, which he explains in terms of uniformity, regularity (*Regelmäßigkeit*) or unexceptionability.[47] Kannisto, in his account of the role of necessity in Kant's theory of causal relations, distinguishes several principles propounded in the literature in relation to the Analogies (the weak, strong, extreme and regulative causal principles) and refers to uniformity as a mark of the strong form, attributed to the interpretations of Lovejoy and Strawson.[48] Watkins, in places in his 2005 *Kant and the Metaphysics of Causality*, refers to the causal powers of the cause, that is, the causality of the cause considered in the Second Analogy, as uniform in nature.[49]

These groups of properties attributed by authoritative scholars might seem to add to the impression that First Analogy provides an infecund environment for temporal determinacy and that, if anything, it aligns more closely with the atemporality dogmas! Consider an eventuality in which the Second Analogy exhibits these properties:

a. The subject causal relations of the Second Analogy are unchanging
b. The subject causal relations of the Second Analogy are both universal and necessary
c. The subject causal relations of the Second Analogy are variously uniform, generalizable, regular and unexceptionable.

Elaborate argumentation is not required in order to convey the impression that these characteristics align on initial inspection with the atemporality conjectures. Not only is it a short logical step from unchanging or generalizable relations to permanent ones; it is tempting to draw from universal, generalizable, uniform such relations conclusions about the validity of time-translation invariance and truth-value constancy. The necessity and universality of relations seem to fit naturally with the constancy of truth values of the propositions describing them, as might uniform relations with invariance through time. The awkwardness of this problem is compounded all the more in the event that the stronger interpretation of the Analogies, albeit perhaps one advanced by a minority of Kant scholars today, is adopted, since this would imply that characteristics (a)-(c) are characteristics of particular empirical laws, and therefore characteristics also of the functional relations addressed by the atemporality conjecture.

However, these conclusions should be resisted. Closer inspection reveals subtle interpretational challenges posed by the Analogies, and blocks direct comparisons between the relations they consider with those of the atemporality conjecture and its sub-doctrines.

Let us make some of these interpretational challenges clear. First, reconsider the conclusions of the First Analogy: time-relations between appearances are determined in relation to the permanence of substance. What is the sense of the term "permanence" in this statement, and is it as comparable to the notions deployed in the atemporality conjectures as might be needed to be used to their advantage? Watkins distinguishes between two immediate interpretations. On the one hand, it could be thought that the permanence of substance endures throughout particular given changes, which is to say that Kant's contention is that for any given time-relation (a given change) some substance exists which is permanent in relation to this change. On the other hand, it could be thought that the permanence of substance demands a stronger interpretation, according to which, for all time-relations (all changes) in appearances, some one substance exists in respect of which these relations are determined. Guyer, for his part, describes this explicitly as a quasi-physical conservation principle.[50] The latter, certainly, would seem to be a natural line of thought given the notable role of the permanent in the Refutation of Idealism, appearing later in the *Critique*, as a necessary condition for the determination of time.[51] Indeed, both in and out of the first *Critique* Kant describes this notion of permanence in more radical terms. The permanent remains always the same,[52] he claims, while, in an unpublished note, he remarks—in seemingly unconstrained fashion—that a necessary condition for the determination of time-relations generally is "something that exists at all times."[53] Kant does not in any telling way relativize the quantifier "all" to a reduced domain such as "all relevant" or "all perceptible," although it is not of course impossible that this limitation was implicit.

There are, however, grounds for doubting whether the best available reconstruction of Kant's matches this rather extreme interpretation. As Watkins points out in his extended analysis of causality in Kant, such would place an onerous burden of proof on Kant: his argument must demonstrate not only that the permanence of substance is required for the determination of changes, but that all such permanents are one, that one permanent determines all such changes, an additional or at least challenging logical step for which little substantiation exists in Kant's writings. Relative, rather than permanent, endurance through change is clearly much more provable. In a similar vein, Allison, in his seminal *Kant's Transcendental Idealism* dismisses the case for taking the permanence of substance in an absolute sense, and considers relatively persisting entities as the conditions of possibility of determinable time-relations.[54] The experience of succession or simultaneity require the persistence of *some* entity, whether or not identical for different succession or simultaneity relations, relative to which they can be discerned, since time cannot be perceived by itself; such relations are experienced as alterations of the states of these entities. Allison refers to these, not as a substratum, but as substrata, underlining the multiplicity of the substances Kant takes to be at work, which in the B-edition becomes the object of the quasi-conservation principle.[55] Alternative commentaries on Kant, such as that of Heidegger, also offer a relative rather than absolute conception of Kant's notion of permanence.[56] To conclude, the permanence of substance is not therefore comparable to the permanence discussed in the atemporality theses. This undermines the initial impression that Kant's Analogies make him an ally of the permanence thesis.

Second, turn back to claims (a)–(c), which suggest that the Analogies also commit Kant to other kinds of atemporality, aside from the permanence discussed in the previous paragraphs. This commitment is, on closer investigation, objectionable. To begin with, the characteristics of uniformity, regularity, necessity and others which occur in these statements are, in Kant, subject to distinctively subtle interpretative issues.

The concept of necessity alone, for instance, is multifarious. Karl Ameriks, in his *On Universality, Necessity, and Law in General in Kant* presents a useful taxonomy of Kant's usage of this notion with more than seven separate distinctions: logical vs. metaphysical, constitutive vs. regulative, universal vs. loose in meaning, among many others.[57] In the *Critique of Judgement*, Kant at times relativizes necessity of judgements to different objects' natures: causes can be causes in an infinite number of ways, each of which has its own law conferring necessity, in addition to the categories' application to objects of experience.[58] Given this complexity, even if it were assumed that the functional relations to which the atemporality theses are addressed admit comparison with the laws presented in their transcendental aspect in the Analogies, whether or not these diverse characteristics could be connected by a credible argument to the specific meanings of the *Critiques* would still not be decided.

In sum, whilst the references in the Analogies to permanence, uniformity, necessity etc. may appear to tether Kant to the metaphysics of atemporality, these appearances are superficial. Additional analysis of these terms undermines this connection. So where does his philosophy really stand in relation to these theses, and what relevance does it have for the temporality of determinacy? Answering this requires attention to be paid to two other aspects of the first *Critique*, namely the unity of apperception and the Schematism.

The Schematism, Constructionalism and the Unity of Apperception

Kant is abundantly clear that a complete understanding of the meaning of the Analogies cannot be achieved by standalone or context-independent analysis of these sections of his works, and that the Analogies depend logically on other principles, by virtue of their status as rules through which their temporal relations to one another are determined.[59] Thus, Kant states that the principle of the Analogies is ultimately grounded in the necessary unity of apperception, which holds in respect of all possible empirical consciousness, and which holds at every time.[60] Indeed, as far as temporal determinacy is concerned, the unity of apperception is germane even independent of this connection to the Analogies. Kant is clear[61] that the doctrine is a necessary condition of avoiding the disintegration of perceptions into rhapsodic lawlessness, and goes as far as to describe it as the foremost principle of human knowledge and the apex of the use of the understanding, and logic.[62] The unity of nature, experience and time are fundamentally related to the unity of apperception—although Kant is not always unambiguous as to the direction of the dependencies at stake, with remarks in the *Reflexionen*, for example, suggesting that the latter doctrine in fact depends for its unity on time as the form of inner sense.[63] Kant defines apperception in several ways:

> Pure, original, unchangeable consciousness.[64]

> The "I think" must be able to accompany all my representations; for otherwise something would be represented in me that could not be thought at all…all manifold of intuition has a necessary relation to the I think in the same subject in which this manifold is to be encountered.[65]

Originally, apperception is defined as that faculty which enables the application of concepts to appearances, and judgements about experience to be made.[66] In addition, Kant distinguishes *pure* from *empirical* apperception. According to the latter, making empirical judgements involves a self-attributing act. Pure apperception thus consists in self-attributing conscious acts. The faculty of apperception, Kant argues, exhibits a unity, which he expresses as the *identity* of the self in all possible representations.[67] Elsewhere, Kant *defines* the understanding as a relation between unity of apperception and the synthesis of imagination, and transcendental apperception this same relation with respect to the transcendental synthesis of the imagination.[68] In Pereboom's reconstruction, several components of this thesis are teased out in greater detail: (i) for any representation, such a representation can be attributed to one's self; (ii) this self is a subject identical for all acts of self-attribution; (iii) such a subject is nonetheless distinct from the representations made thereby; (iv) such a subject is capable of being conscious of its representations.[69] Strawson, for his part, emphasises the self-attribution of mental states as the doctrine's distinctive element, whereas Allison emphasises the *possibility* (rather than the actuality) of accompanying acts of thought for all representations.[70]

What insight does this doctrine generate for temporal determinacy? Kant describes the unity of apperception as a necessary condition for experience, providing for the various different syntheses brought to bear on appearances in the application of concepts. It is in and through synthesis that apperception provides *rules* which apply to appearances and confer determinacy on them. To the extent that the work of the categories in organising appearances, such as was seen in the Analogies as well as in Kant's Transcendental Deduction, is rooted in the unity of apperception, so too can meaningful conclusions be drawn as to the temporality of determinacy from an analysis of this facility. This is the meaning of the thesis that the synthetic unity of appearances is necessary for experience, facilitated by the connection of possible consciousness which defines the unity of apperception. Whilst Kant's presentation is certainly protean, it contains one especially salient datum for this analysis, namely the claim that apperception is an *activity*.

In the *Critique of Pure Reason*, Kant defines pure apperception as an act of the mind. In remarks on a letter from D. F. von Lossow Kant emphasizes what he describes as an activity (*actibus*) of thinking and presents this as a definitive aspect of the occurrence of objective experience via apperception.[71] It is an action of the mind—an action of determining and determination—a framing Kant deploys several times—which enables successive experience to be constituted. Apperception consists in laws of the mind's activity,[72] Kant claims, as he relates these laws to the abrogation of contingency in nature. Moreover, the active nature of the mind in its relation to appearance is visible also in the very way in which Kant defines the understanding: whereas the mind's receptivity is *defined* as sensibility, understanding can be defined as the *spontaneity* of knowledge.[73] Wolff, in his meticulous *Kant's Theory of Mental Activity*, talks of Kant's identification of the faculty of the understanding with its activity, and paraphrases Kant's analysis of the synthesis effected by apperception as one which casts the synthesis as a *product* of the understanding—such products being the outcome of a process effected by activity.[74] Wolff later refers explicitly to synthesis as such a process.[75] Allison, in his analysis of the process of empirical concept formation, characterises these as constructed by the understanding's acts, and points to Kant's claim that such acts are rooted in the spontaneity of thought.[76] In turn, Pippin, in his essay "Kant on the Spontaneity of Mind,"[77] specifically associates this spontaneity with the acts and inner determinations of apperception.[78] Importantly, (*contra* Sellars) Pippin takes this act of spontaneity as absolute *qua* indeterminate, and irreducible to a causal relatum.

The upshot of this is the following. If apperception confers rules on appearances, and the imposition of such rules is itself an act, and such acts are themselves indeterminate acts, the rules conferred by apperception would be brought about by such acts. Pippin's position in this respect is no idiosyncrasy—Watkins opens up similar questions as to the determinate or indeterminate character of the apperceptive activity which confers determinacy on appearances and, by extension, the causality of the causes which arise from this activity. Watkins considers the causality of these causes to be indeterminate: he goes on to declare explicitly that temporal determinacy can only be brought about by an act by which it determines certain states and therefore that the causal efficacy of the cause, that which is causal in the cause, is not itself determinate.[79] It is in this way that activity of the mind thus serves as the undergirding of change in appearances.

There are therefore several crucial theses which qualify apperception as an active facility.

> This synthetic unity in the time-relations of all perceptions, as thus determined *a priori*, is the law, that all empirical time-determinations must stand under rules of universal time-determination. (B220)

> This synthetic unity is the (*a priori* and transcendental) unity of apperception. (B220)

This apperception is an activity of the mind and an action of determining.

The rules of universal time-determination under which stand empirical time-determinations are the activity of the mind.

Now, this connection between rules of time-determinations and the activity involved in apperception means that assuming this activity of the mind and action of determining to be spontaneous, irreducible, indeterminate in the manner described variously by the likes of Watkins, Pippin, Wolff, and others, imparts these properties to the former, too. Of course, the rules of universal time-determinations and the empirical laws of nature which hold of appearances are in no way coterminous: just as Kant's references to the permanence, universality and other characteristics of the rules of the *Analogies* need not entail the atemporality theses which concern functional relations, a cleft remains between the rules produced by the synthetic activity of the Kantian mind and the laws whose temporality has been in question in previous chapters of this monograph. But these remarks at least bring into reach a basic connection between the two, and highlight the import of Kant's work for temporal determinacy, as well as its coherence with a philosophy which affirms the universality and necessity of laws.

Before drawing further conclusions as to the nature of the activity which, through apperception, impinges upon determinations of time, it is instructive to refer to another of the three obvious aspects of the *Critique of Pure Reason* in which this notion is expounded and developed. The unity of apperception is not the only area of the *Critique*, not to mention other texts, in which the determinacy of nature is constructed by the activity of the mind. Kant's "Schematism of the Pure Concepts of the Understanding" takes up the role of schemata, being for Kant determinations of time in conformity with rules, pure time determinations, enabling the representation of determinations of time.[80] The schemata mediate between objects and concepts, and allow Kant to reconcile his claim that sensibility and understanding are, in some meaningful sense, separable, and that objects and concepts are heterogeneous, with his claim that the understanding's concepts nonetheless *subsume* objects—a problem sure to arise for any attempt to ground our experience of one nature in two distinguishable faculties. Kant refers to the schematism of the categories as an activity, and to the schemata themselves as its products; the schema of a concept is a pure synthesis, and a transcendental product of the faculty of the imagination.[81] The schemata establish in and through the activity of the mind certain time-determinations relating, according to Kant, to the series, content and order of time.

Now, several influential interpreters present the schematism as integrally active in its relation to rules of appearances. It is worth exploring these briefly before defining their key commitment—the Constructionalism Thesis. To begin with, Eva Schaper, in her essay "Kant's Schematism Reconsidered," defines and analyses in detail what she terms a constructionalist interpretation of this aspect of Kant's work, defined by the precept that the synthesis which effects the unity of experience is constructed by the mind.[82] Schaper advances two distinctive propositions which frame her interpretation. First of all, the productive imagination is the primary vehicle for the schematism, which admits only a secondary relationship to the understanding, reason and sensibility. Second of all, the schemata themselves are pure time-determinations.[83] Schaper goes onto draw attention to the liberal employment of participles in Kant's presentation of the dynamics of synthesis, that is, the form*ing*, shap*ing*, structur*ing* of perception by the imagination.[84] Accordingly, she describes how the imagination *moulds* the manifold of intuition in order to enable its subsumption under concepts (overcoming the problem Kant identifies at the beginning of the Schematism, namely, how one of two heterogeneous representations can nonetheless subsume the other). Indeed, at times, such as in his correspondence with Beck, Kant intimates that space and time themselves, rather than determinate objects given in them, are subject to this constructive activity; the mind must undergo a process of composition or of composing objects and even forms such as space and time.[85] Thus, Schaper claims, this subsumption is ultimately a contribution of and derives from subjective activity. Her much more general conclusion is that the span of human experience consists in different modes of temporal construction.[86] Most importantly for present purposes, these statements amount to a definition of the constructionalist interpretation of the schematism.[87]

A number of scholars defend positions which conform to Schaper's definition of constructionalism. Samantha Matherne, in her "Kant and the Art of Schematism" proposes a break with the tendency to obscurantize or mystify Kant's schematism, a tendency she locates in the work of luminaries of Kant studies as Strawson, Bennett and Guyer, all of whom wield Kant's famous reference to the schematism as a concealed art, bound up with the mind's deepest inner movements, as his own acknowledgement of having failed to fully reach a lucid formulation of the doctrine.[88] This impression is certainly an easy one to come to endorse when one takes some of the key passages of Kant's, such as A139/B178, in which he passes in only a few short sentences through a suite of subtle concepts—the synthesis of the manifold, time as the form of inner sense and the determination of time—to infer the homogeneity of a schema with both categories and appearances. Matherne, by contrast, advocates taking literally Kant's description of the schematism as an art. This interpretation also defines the schematism as an activity (*Kunst*):[89] the activity involved in the schematism of the categories is a manifestation of the dynamic capacity of the imagination which is indispensable for cognition as much as for morality and aesthetics. The imagination, defined by Kant as the faculty which enables non-present objects to be represented, can be glossed as a clutch mechanism between sensibility and understanding, and enabling the synthesis rooted in the unity of apperception to work on actual appearances.[90] Not unlike Pippin, Matherne repeatedly describes these aspects of the imagination's activity as spontaneous in nature.[91]

Finally, this catalogue of constructionalist thinking about Kant's schematism would hardly be complete without mention of Heidegger. Heidegger's study *Kant and the Problem of Metaphysics* orbits around the imagination and its role in the Schematism—as he characterises it, the meaning of the becoming-sensible of concepts.[92] Heidegger views these passages as the essence of the first *Critique*, and advances a number of theses which align him with thinkers of the activity of the Schematism. The Schematism itself is considered to be something which *occurs* and which *springs forth* from the imagination.[93] Elsewhere, Heidegger even locates the provenance of time itself in the transcendental imagination.[94] Already, these terms are surely marks of a process rather than an atemporal state. As will be seen shortly, Heidegger's analysis of the ground of the Schematism, which lies ultimately in the transcendental imagination and inner sense, dismisses not only those commentators who take these as facilities of a substantial subject, but objects to the characterization of the facilities themselves as unchanging, eternal, or perduring—a line of thought which also strengthens the link between constructionalism and temporal determinacy.

This account of the grounds of time-determination in the *Critique of Pure Reason* can be summarised as follows: both Kant's account of the role of apperception, and in particular the doctrine of the unity of apperception, and of the nature of the schematism, present the mind's relevant facilities as active facilities. From the standpoint of constructionalism, these facilities enact events and processes as opposed to stationary rules. How can this position be stated more precisely? A framework developed by Patricia Kitcher is particularly useful for this purpose. She considers the objects subject to the schematism in terms of states, and provides a formal expression of the pure synthesis which applies to them. For two states S_1 and S_2 to be synthesised is for it to be the case that:

Given two states S_1 and S_2 there is a further state S_3 such that S_3 would either (i) not exist or (ii) be different, had one or both of S_1 or S_2 either (i) not existed or (ii) been different.[95]

This schema provides a clear and general framing of empirical synthesis, namely the synthesis of delineable states, such as those given in intuition, like visions or sounds of this or that object. Kant's more elusive notion of transcendental synthesis has been taken by scholars as a synthesis which is a condition of possibility of experience, or as a constraint which empirical syntheses must obey.[96] Formulated in accordance with this schema, constructionalism can be cast as the claim that any such synthesis S_3 is an activity, and that occurrences of the term "state" in no way suggest this synthesis is static or atemporal.

Given two states S_1 and S_2 there is a further state S_3 such that S_3 would either (i) not exist or (ii) be different, had one or both of S_1 or S_2 either (i) not existed or (ii) been different, and S_3 is an activity (Constructionalism Thesis).[97]

This is the thesis which defines constructionalism. What implications, if any, does this have for the temporality of determinacy? The claim that synthetic activity enables temporal determinacy would be the claim that S_3 itself depends on some other parameter τ which varies throughout a process. Such a parameter would have to be distinguished from the primary notion of time Kant deploys explicitly as form of intuition, formal intuition, and form of inner sense.[98] It would have nothing to do with role of time in the Transcendental Aesthetic. Before moving forward, then, it is necessary to investigate whether this distinction is coherent, or whether it precludes any use of Kantian constructionalism in the service of this chapter's goal.

Ralph Bader considers explicitly the distinction between these two aspects of time in his "Inner Sense and Time." Whilst time is primarily involved in inner appearances as the form or framework in which these appearances are represented, Bader also proposes the existence of an alternative time in which these representations themselves are made and come to pass. Whilst Kant rarely acknowledges the legitimacy of the latter, such is needed, for Bader, in order to account coherently for present representations of temporally disparate content, that is, representations which are present, but which represent contents as occurring at different times. Whereas outer (spatial) intuitions involve the representation of content ordered in space, inner (temporal) intuitions involve a present inner representation, consisting of ordered representations. Bader characterises the latter as a noumenal representation which contains a phenomenal representation of representations—thus, a representation of a representation of representings, or acts of representation.[99] Memories, for example, involve the present representation of appearances represented as occurring at different times. One represents prior acts of representation, not merely prior representations: memories are memories of actions of inner sense. M. Valaris, in his "Inner Sense, Self-Affection and Temporal Consciousness" argues for a similar conclusion. He takes the temporality of inner states as a component of the most credible account of Kant's distinction between inner and outer temporality—that is, the temporality of inner sense, and the temporality of intuited outer appearances. According to Valaris, the subject's self-affection is itself subject to a temporal order.[100] Subjective states, acts of self-affection of the subject, constitute a radically interior time of their own, and a form distinct from the form in which representations are intuited. He points to textual evidence in support of this view, such as the introduction to the *Analogies*, where Kant refers to the mysteriously *contingent* way in which acts of inner sense juxtapose empirical intuitions,[101] as well as his account of the relation between inner and outer sense. These instances of self-affection analysed by Valaris, in turn, are identified with acts of figurative synthesis which, as above, impart determinacy to sensibility.[102] If such acts of synthesis which were represented before with the symbol S_3 are equally acts of self-affection, and acts of self-affection admit a distinctive temporal order, a temporality of the determinacy endowed by synthesis would appear to have been constituted.

On the face of it, positing *another* time in which the activity of synthesis occurs would appear to generate inconsistencies with the letter of Kant's writings. For one, Kant clearly states that time is unified, one and unchanging and, in his comments on a letter from Lossow, as unitary, infinite, necessary and a priori.[103] He refers to apperception as an unchangeable form of consciousness.[104] Specific schemas, such as that of substance, are explained in terms of persistence in time.[105] In an unpublished note from the 1770s, Kant defines time as the condition under which the subject becomes aware or conscious of things.[106] In a note composed during the visit to Konigsberg of Kant's student and proof-reader J. Kieserwetter, Kant considers the status of what Bader refers to as the phenomenal representation of representations, namely the way in which existence is represented as empirically determined—that is, the phenomenal self, intuited in time analogously to any other sensible content. In so doing, he appears to dismiss the possibility of a multiplication of times. Specifically, Kant's refutation takes two forms. Firstly, the act of consciousness responsible for time-determination cannot be represented as contained under the conditions of time-determination of one's state. Secondly, time, as the form of inner sense, cannot be *contained under* another time. His argument consists in this linked series of statements:

> The consciousness when I **institute an** experience is the representation of my existence insofar as it is empirically determined, i.e., in time. Now if this consciousness were itself in turn empirical, then this temporal determination, as contained under the conditions of the temporal determination of my state, would in turn have to be represented. Yet another time would therefore have to be given, **under** which (not in which) the time that constitutes the formal condition of my inner experience would be contained. Thus there would be a time in which and with which at the same time a given time flows, which is absurd.[107]

Yet temporal determinacy and the distinctions made by this present chapter, fall within the scope of neither of these two infractions. The temporal determination of *inner sense itself* is not contained under the conditions of time-determination of one's *state*. The conditions of time-determination of one's state are the conditions which govern intuitions about one's self—one's feelings, instincts, and perceptions of self. The latter are contained under the same time as are empirical intuitions, not the time to which acts of synthesis are tributary. Nor by the same token is this latter time, in Kant's terms, contained under another time which governs acts of synthesis; it is this time which so governs these acts. Indeed, it is clearer from the sentence following this passage that the repudiation Kant has in mind is one directed principally at the thesis that consciousness itself is empirical, or an object of experience, rather than a condition of possibility for objects of experience, which is to say, transcendental in nature. By denying that consciousness could itself be empirically determined, Kant pre-empts the Humean counter-thesis which might attempt to reduce acts of apperception to associations of perceptions; Kant anticipates that such would lead to a pernicious regress in which each such reduction only ends up requiring a further transcendental condition for its determination. Only if one assumes that acts of synthesis are themselves parts of empirical consciousness, comparable to representations of one's physical person, would the time which governs said acts govern empirical representations, and therefore require a further time for its own determination. By contrast, this condition does *not* obtain for the constructionalism-inspired interpretation of synthesis and apperception described in this section.

What of, however, the other claims Kant makes about time, in his better known metaphysical works? In spite of what Valaris, Bader and others claim, are these not harder to reconcile with the claims that time is (i) unified and (ii) unchanging and (iii) perduring, as we saw Kant propose in the first *Critique* as well as in other texts? Of course, decisions as to this question depend on decisions about the meaning of these three terms. If constructionalism with respect to synthesis (and apperception by extension) is to be a viable point of entry of temporal determinacy into Kant's thinking then the frequent references Kant makes to these characteristics of time throughout his opus must be explained as a priority.

As respects constructionalist and related approaches, Heidegger's analysis in *Kant and the Problem of Metaphysics* deals extensively with the meaning of these characteristics, reconciling them with his vision of Kant's philosophy of time. As intimated above, Heidegger's account emphasises the centrality of the schematism in the *Critique*, and the imagination as the schematism's determinative faculty, speaking of it as an event, an occurrence, and a moment of springing-forth—emphases all of which attest to the active nature of the mind's synthetic faculties, in accordance with the Constructionalism Thesis.[108] In his account, Heidegger undertakes a subtle close-reading of Kant's references to time as fulfilling such properties as (i)-(iii) and attempts to define them in detail, paying attention to the fact that these properties have a rather different signification when employed in the unique context of the schematism over against their everyday meaning. To begin, the claim that time is unchanging or permanent, for Heidegger, states nothing more than a relation of identity between synthesis and itself. To deem time unchanging is thus to deny that time becomes other than itself, and in this sense that time passes. Time is *always now*, as it is now *in every now*, and is *permanent* in this sense. For Heidegger, then, the immutability of time and the fact that time does not pass both equate to being now in every now.[109]

How can one make sense of this claim? After all, defining the permanence of time is here in no way comparable to defining a property or attribute of an event, state or substance. To what kind of time is Heidegger referring? Heidegger implicitly appeals to an important distinction between two types of time-relations. On the one hand, time-relations between events, such as succession or simultaneity, are attributed to intuitions, contents of the manifold, just as sounds of thunder follow lightning, and visions of dark clouds are had with the illumination of the sky. These relations are the focus of the arguments of the Transcendental Aesthetic in which the conditions of possibility of these relations are located in the a priority of time, as a form of intuition, independent of experience, prior to experience, and belonging to the constitution of the mind. Changes observed in nature are constituted in these relations: here at some earlier time some data are given simultaneously; there at a later time some other data succeed them. On the other hand and by contrast, the time of which (i)-(iii) are predicated is not an object which admits such relations as these, but is rather their ground. For example, Heidegger claims that time is not to be taken as a thing among other things.[110] The relevant relation in the case of time is neither the succession nor the simultaneity of appearances, but rather the relation of an act of inner sense to the material synthesised by this act. Even Heidegger's allusion to sequences of moments is misleading in this aspect: acts of inner sense are not comparable to moments which can be dimensionalized by arranging one alongside the other, as if neighbouring houses in a row. The consciousness in which an experience is instituted, the spontaneity of thought, in Kant's terms, is the same regardless of whatever empirical content stands under them, for such acts, being pure, are independent of the ultimate objects subject to synthesis by the imagination. It is permanent and unchanging in the sense that each act is the same, not in the sense that they admit the same content as they pass through time, as a statue might endure years of adverse weather. The second-order time in which synthesis occurs is one time, insofar as there is no passage from moment to moment, nor any higher synthetic faculty which could retain or anticipate other acts of synthesis against which a given act could be contrasted: synthesis is the same for each and every act. Thus, Heidegger refers to the transcendental power of imagination as one self-same spontaneity[111] from which time springs forth.[112] This perspective reinforces Bader's denial that the temporality of inner sense implies that different acts of synthesis are contained under another temporal dimension—as would be ordinary events subject to temporality—shielding this approach from the accusations of absurdity Kant directs at empiricism with respect to time. Further, such a perspective would have nothing to do with Kant's various assertions as to the static nature of time as form of intuition and the view that the time in which changes of appearances are thought remains and does not change. Heidegger therefore claims to have no need to efface Kant's commitment to (i)-(iii). On his view, the fact that any and all acts of synthesis which determine time are the same act of the mind, and the fact that these are independent of any empirical content which could change from act to act establishes the unity of time, unchanging, permanent and perduring.[113]

We have several conclusions. These bear repeating from the very start in order not to lose track of the argument of this chapter. There are three obvious aspects of Kant's critical works with immediate relevance to the temporality of determinacy. In the Analogies, Kant considers directly relationships of causal determination applying to appearances, and in ways which seems at first to cohere atemporality thesis. However, the uncertain link between these causal relationships and the functional relations of the atemporality conjecture undermine this comparison, and force the reader to distinguish different senses of the term "determinacy" operative in these distinct contexts. In order to understand exactly what kind of determinacy is at stake in the Analogies, and its origins, the doctrine of apperception and the transcendental schematism must also be considered: they are responsible, Kant says, for the time-determination of the manifold, for the imposition of rules on appearances. According to an interpretation of this theory known as constructionalism, synthesis is an activity, rather than an abstract rule independent of the subject's cognitive faculties—an interpretation which can be summarised by the following Constructionalism Thesis:

Given two states S_1 and S_2 there is a further state S_3 such that S_3 would either (i) not exist or (ii) be different, had one or both of S_1 or S_2 either (i) not existed or (ii) been different, and S_3 is an activity (Constructionalism Thesis).[114]

This thesis describes Kant's extremely frequent references to activity in the context of his analyses of synthesis and apperception. Despite occasional appearances to the contrary, interpretative resources are available which reconcile this interpretation to Kant's remarks as to permanence, unity, unchangeability, and perdurance, with aspects of the work of authors such as Hegel, Heidegger, Bader, Kitcher and Valaris all contributing to link between these characteristics and the singularity of constructive activity.

In order to finally secure the relationship to the temporality of determinacy of this constructionalist interpretation of the *Critique of Pure Reason* and ancillary tracts, a further step is needed. The Constructionalism Thesis (CT) can be linked to temporality, by virtue of the fact that activity, particularly the spontaneous activity considered by Kant, is by nature processual, and in this sense temporal. Cast in the terms of the CT, S_3 is a process as well as an activity. Thus, Petersen characterises the activity of the imagination as a process of making thought sensible; Watkins describes such processes as being continuous in their efficacy; Heidegger refers to regulating processes as being rooted in the schematism; Dahlstrom, in his review "The *Critique of Pure Reason* and Continental Philosophy: Heidegger's Interpretation of Transcendental Imagination" speaks repeatedly of the process of holding together experiential contents in an imaginative synthesis.[115] Schaper goes so far as to suggest that well known twentieth Century philosophical systems such as those of A. N. Whitehead and Henri Bergson were influenced to a significant degree by the subject matter of the schematism, and that these aspects of Kant's metaphysics serve as something of a conceptual seedbed for these philosophies of process. More directly, Kant, in B158 in the first *Critique*, claims that, prior to the act of the mind which confers determinacy, no self-intuition—no consciousness of self—is given:

> Now since I do not have another self-intuition which gives the determining in me (I am conscious only of the spontaneity of it) prior to the act of determination, as time does in the case of the determinable, I cannot determine my existence as that of a self-active being; all that I can do is to represent to myself the spontaneity of my thought, that is, of the determination.[116]

Consider, then, the label τ_0 which describes the time—not time *qua* form of intuition, nor the time of empirical appearances, but rather the parameter proper to the process of synthesis—at which this act occurs. For $\tau < \tau_0$ the implication is clear enough from this extract that there is no such determining activity, and that determination has not been conferred, remembering here that this determining activity enacts the spontaneous synthesis of the manifold, the imposition of rules on the manifold, and the time-determination of appearances, just as Kant equates spontaneity of thought with determination. Once a partition in the set $\{\ldots\tau_0, \tau_1, \tau_2, \ldots\}$ is introduced, once a distinction is made between these "times" at which determinacy has been enacted and those others at which it has not—then an atemporal conception of the subjection of the manifold to rules can no longer be upheld. If the synthetic activity of the mind is established as a condition of possibility of the determination of appearances, once this activity is established as a process which does and does not occur then and now, the temporality of determinacy enters into the Kantian discourse.

Systematicity and Purposiveness

The processual nature of the activity of the mind is no singular oddity of the *Critique of Pure Reason* and Kant's notes surrounding it. Nor therefore is it the only evidence which exists in Kant's writings for temporal determinacy. In the *Critique of Judgement*, a completely new perspective is advanced on the working of the understanding and, in particular, the process of judgement, and one which is in fact more integrally connected with the laws of nature. A number of scholars including Ginsborg, Kitcher, Rescher, and Guyer have developed interpretations according to which systematicity is a necessary condition for the lawfulness of nature. The systematicity which holds of the laws of nature is itself a product of the activity of the mind, in its construction of coherent theories from individual empirical regularity: even in his earlier critical texts, such as in the *Critique of Pure Reason*—"Regulative Employment of the Ideas of Pure Reason"—Kant refers clearly to the rule and principle of systematic unity as resulting from a procedure of the understanding and, again, the activity of reason.[117] This approach has several implications, including the quantitative indeterminacy of the systematicity constituted by the use of judgement, the multiplicity of laws and natures described by such laws consistent with this use, and the time-dependency of the systematization so effected.

The rationale for these conclusions can be discovered by reviewing the work of several prominent commentators on Kant's conception of systematicity and its relationship to laws.

1. Philip Kitcher, in his "Projecting the Order of Nature", attributes several claims to Kant. First of all, a given law *L* constitutes a law of nature if and only if *L* plays a specified role in some system of beliefs

S constructed in accordance with a set of rules {*M*}. Further, causal relations are considered to hold between different types or kinds to the extent that such relations play an analogous role in *S*. Second of all, such laws of nature are considered to be *necessary* to the extent that they play this role. Thirdly—and, *prima facie*, startlingly given this attribution—Kant also unequivocally characterises such laws as contingent in a number of places: in his discussion of teleological judgement, for example, he considers the example of a bird much of whose anatomy, from its wings to its tail, could have been arranged differently.[118] Mysteriously, Kant describes how the universal laws of nature which belong to the understanding are needed in order to constitute objects of experience, and how particular empirical laws are contingent, but somehow nonetheless *regarded* as necessary by the understanding in order to form a single order of nature.[119] Elsewhere in the Introductory sections of the *Critique of Judgement*, Kant also claims that out of principle laws must be *taken as* necessary, though not that they are necessary, drawing attention again to the cleft between the truth of propositions within his own transcendental-critical system about the way things are represented, and the verisimilitude of these representations.[120] Such a cleft is visible also in the *Prolegomena*, in which he speaks of the validity of the synthetic unity of perceptions being *represented* as universal and necessary.[121] This allows a reconciliation of sorts between two seemingly incompossible conditions, not so much simultaneous contingency and necessity of any rule or concept, but the *de facto* contingency of empirical laws together with the transcendental attribution of necessity to them by the understanding.

2. In this vein, Messina frames Kitcher's view as one according to which systematization *requires* the belief in the necessity of some laws, rather than one which independently establishes said necessity. Nonetheless, as Kitcher notes, this does not fully discharge the tensions built up in Kant's multi-faceted account of systematicity. There remains a paradoxical sense in which such laws, despite the strength of the claims of universality and necessity attached to them in his earlier writings, lack objective validity. Kitcher notes the obvious antagonism involved in these claims: how could contingent laws represented as necessary in

accordance with a certain concept nonetheless fail to be objectively valid?[122] This is the foundation for his conclusion that the order of nature, the unification of experience, is projected by the activity of the mind.[123]

3. Guyer provides a detailed analysis of the Kantian notion of systematicity in his paper "Kant's Conception of Empirical Law,"[124] pointing to Kant's imputation of objective validity and necessity to what he describes as nature's systematic unity in the final stages of the first *Critique.*[125] A number of noteworthy characteristics, Guyer observes, appear in the *Critique of Judgement* and *Opus Postumum*: systematicity is described variously as asymptotic, indeterminate, approximate, and yet no less a transcendental and necessary principle—that is, for Kant, one which enables representation of the conditions of possibility for things to become objects of cognition.[126] Kant compounds this seeming antinomy by harbouring what appears to be a conception of the multiplicity of determination brought about by laws. As well as the universal, necessary laws which result from the application of the categories to the formal conditions of intuition, Kant hints at the existence of objects of cognition which are determined and also determinable in a number of ways as well as what he describes as differentiable natures which can be causes in an infinite number of ways.[127] Consequently, as respects particular empirical laws, i.e. the functional relations which make up scientific theories, many different combinations can determine any object. This makes it puzzling how it could be that any *one* theory, as an individuated set of empirical laws, could come to be adopted, and that at the same time this theory could be seen as necessary. As a result of his attempts to clarify the mysterious and ambivalent status of the principle of systematicity, Kant is forced to distinguish between two types of principles: constitutive and regulative. Whereas regulative principles determine nothing in objects and determine what is rational, constitutive principles determine objects, and thus what is real.[128] As a regulative principle, systematicity amounts to an ideal to which our conception of laws must conform in order for us to be able to think nature as at all unified, and thus transcendental, albeit not objective. As for Guyer's guiding question as to the logical relationship between the tenet that nature is

systematically unified and the validity of empirical laws, the former is construed as a transcendental condition, condition of possibility of the latter.

4. Teufel, in his analysis "Kant's Transcendental Purposiveness" and the "Maxim of the Lawfulness of Empirical Laws," echoes these claims. Once again, Kant's aversion to the possibility of disorder and disunity in cognition leads him to posit systematicity of nature as a necessary condition for the epistemically cogent theorization about nature. Teufel claims that to endorse this principle of systematicity is to eschew the possibility of chaos in appearances since, without it, our cognitions of the world would be transient, ephemeral, clashing and contradictory. Accordingly, Teufel's conception of systematicity is not so much the objective cognition of empirical laws as necessary, but rather the entitlement of the understanding to cognizing them as necessary in spite of knowing their contingency—a metaphysical rather than logical or empirical necessity. This notion of systematicity Teufel equates with one of purposiveness, defined by Kant in the Introduction to the *Critique of Judgement* as a concept by virtue of which:

> ...nature is represented as if an understanding contained the ground of the unity of the manifold of its empirical laws. The purposiveness of nature is, therefore, a particular *a priori* concept, which has its origin solely in reflective judgement. For we cannot ascribe to the products of nature anything like a reference of nature in them to ends, but we can only make use of this concept to reflect upon them in respect of the nexus of phenomena in nature—a nexus given according to empirical laws.[129]

Kant also refers to this principle as one of reflective judgement rather than determining judgement, a mode in which one reflects on the unity of nature, attributing nothing to the object.[130] Thus, by virtue of this concept, nature is represented as if the unity attributed to nature originated in the understanding. Accordingly, like Guyer and Walker, Teufel holds the principle to be a technical and practical principle—not a constitutive principle of the understanding nor one which attributes anything to objects—and one which holds asymptotically.[131]

Teufel is explicit that any cognition which takes place before reaching this asymptote at infinity must be revisable and iterative.

According to this review of the literature, then, there are several propositions that can be attributed to Kant whose conjunction in the *Critique of Judgement* expresses what might be called his principle of systematicity, or principle of the purposiveness of nature:

A given law *L* constitutes a law of nature if and only if *L* performs a specified role *R* in some system of beliefs *S* constructed in accordance with a set of rules {*M*}
L is represented as necessary if and only if it plays this role
L is contingent
The role *R* in *S* in accordance with rules {*M*} is performed in a way which is asymptotic, indeterminate, and approximate.
1. is a transcendental principle.[132]

Kant's theory of systematicity therefore has several components. It sets out a coherentist necessary and sufficient condition for a given law to count as a law of nature. For Kant, laws are, in part, laws by virtue of rules which specify that they qualify the theory in which they appear as providing a coherently unified and systematic account of nature. Of course, this does not preclude their playing other roles, such as the way in which they confer determinacy on the events they describe; Kant's objective here is not to produce an exclusive or exhaustive definition of laws. Further, this theory characterises laws as being *represented* by these rules as necessary if they confer such a qualification. Nonetheless, the laws themselves are contingent. Accounting for the obvious dissonance between these two properties requires Kant to distinguish between two different uses of the understanding: over against the transcendentally necessary principles of, for example, the *Critique of Pure Reason's* Postulates of Empirical Thought, the use of the understanding in the integration of laws into a system of beliefs is regulative rather than constitutive. As such, this use is asymptotic, indeterminate and approximate.

Before exploring the connections between this principle of systematicity or purposiveness and the overall temporal status of laws, a further observation must be made which concerns the way in which the understanding carries out its regulative function. The approximate, indeterminate, asymptotic role played by laws in theoretical systems, provides a quite different line of sight on the faculty of understanding to that emphasised in much of Kant's earlier critical works. As Guyer observes, the principle of systematicity is not so much prescribed to objects but rather enacted by the understanding in a procedure. Kant is clear that nothing is determined in the objects by the principle of systematicity, but rather the activity of the understanding itself. The understanding conforms to the principle through a process in which it actively works on the belief system (*S*). Such work is as iterative and as processual as its result—the systematicity of laws—is approximate and indeterminate: an indeterminate yet harmonious activity.[133] Kant's framing in the Introduction to the third *Critique* makes it abundantly clear that a process and a journey is involved in the utilization of the principle of purposiveness. The extended labours involved in searching for laws, whose heterogeneity preordains the discovery of laws inconsistent with those in any previously accepted theory, and the existence of others which are undiscovered, are guaranteed by the fact that no definite limits can be set on the use of the principles in organising empirical laws.[134] It is not that nature must be assumed to be systematic and ordered, and that chaos and disunity must be eschewed, in order for the understanding to constitute a static, rule-governed manifold, but rather that this assumption is needed to develop, in a progressive way, a proleptic theory of the world. Such order is in this sense the result of intellectual activity, and an open-ended process of projection.[135]

There are, therefore, two senses in which the Kantian *oeuvre* conceives the mind's work as a process and an activity. First of all, it has been shown that a credible interpretation of crucial elements of the *Critique of Pure Reason* and texts which illuminate it—the Analogies, the Schematism, and the Unity of Apperception—holds that synthesis can be described as an activity and a process carried out by the mind:

> Given two states S_1 and S_2 there is a further state S_3 such that S_3 would either (i) not exist or (ii) be different, had one or both of S_1 or S_2 either (i) not existed or (ii) been different, and S_3 is an activity (Constructionalism Thesis).[136]

Secondly, it has been shown that the later Kant's reflections on purposiveness and systematicity, especially in the *Critique of Judgement*, posit these as transcendentally necessary aspects of the way in which laws describe nature. The laws in question must be regarded as necessary in order for this description to describe anything coherent at all, and yet they are formed and reformed in a sempiternal process which may never be complete, again as a result of the mind's activity.

Equipped with these conclusions, little further analysis is needed in order to ascertain the status of temporal determinacy. First of all, the CT impinges upon temporal determinacy in a limited and specific sense. The synthesis S_3 is repeatedly framed by Kant as the vehicle which imparts determinacy to appearances: in order that something might happen, it must be determined in accordance with a rule imparted by the categories of the understanding. Such categories in turn amount to acts of synthesis as Kant suggests in, for example, the Second Analogy.[137] Now, it has been shown that insofar as S_3 is an activity, and this activity is a process, such determination is imparted by virtue of a process, or progressively.[138] Consequently, if such processes are in some legitimate sense temporal, the activity of synthesis and the determinacy it imparts follow along. Kant is clear that events occur by virtue of their relations to other events: every occurrence presupposes a rule to which it is subject and all alterations conform to law.[139] These relations determine them objectively to occur, and to the properties they instantiate. Being the product of the activity involved in synthesis, they are temporal in a similar sense as is such activity.

Hannah Ginsborg, in her study of the peculiarities of Kant's "Lawfulness without a Law: Kant on the Free Play of Imagination and Understanding" outlines two key senses in which this result temporalizes the determinacy brought about by these rules. First of all, the determinacy conferred by these rules cannot be grasped antecedently to the activity on which they rest. Specifically, Ginsborg alters the relation conceived by a number of Kant scholars to obtain between synthesis and rules: synthesis, for Ginsborg, *exemplifies* rules.[140] There is no preceding conception of which events should or should not occur to which appearances synthesised must conform. Rather, they are determined concurrent to the activity of synthesis. Just as a student of a language demonstrates their grasp of lexical and grammatical rules as they use and apply them, and as the legitimate and illegitimate instances and usages of these determine the rules to which they are subject, so events related to one another by rules are related as acts of synthesis occur. Thus, for Ginsborg, rules take shape in the activity they govern, and are no less rules on that account; it in no way follows from the nature of the rule that it must pre-exist the exercise of synthesis. In this sense, the rules are internal to the activity. The description "rule-governed" does not imply the description "determined in advance."[141] Second of all and relatedly, Ginsborg's model holds synthesis to be a free activity. Insofar as rules are not imposed by the activity of the mind as if a seal imposed on wax by a stamp, but are rather constituted by this activity, activity may be characterised as free in relation to such rules. Since she maintains that the rules in question are no less regular and lawful on account of this fact, the activity of synthesis becomes a matter of free lawfulness, lawfulness without a law, or indeterminate lawfulness.[142] In these two specific senses, Kantian determinacy is subject to time.[143]

However, whilst this line of thought enables some initial purchase to be found for temporal determinacy within the Kantian architecture, it has a limited and technical remit. The question of the temporality of determinacy in this work is addressed by confining the meaning of the two terms "temporality" and "determinacy" in a manageable way, taking determinacy as the determination of an event imparted by a rule instantiated in a *functional relation*, and using the atemporality theses, which contemplate translations in time and the permanence or variability of laws, to be specific about what kind of temporality is in question. The meaning of "determinacy" in Kant's writings is, however, distinct and restricted, and motivated by very specific Kantian problems. This has already been discussed in the above assessment of the implications of the First Analogy, with the question as to whether its references to permanence pre-empt Kant's perspective on temporal determinacy.

The functional relations of the atemporality theses cannot be said to be directly contemplated by the properties attributed to the synthetic cognitive activity. Whilst they bear on the temporality of determinacy in the sense just outlined, it cannot be said that the functions *f* whose truth values do or do not vary over time, or which are said to be free of variables such as *t*. As a result, extending the sense argued for in the passages above to the characteristics of functional relations, and using them to test, *a fortiori* to refute, atemporality conjectures—this would not be possible as this argument stands.

However, this analysis considers only the *first* respect in which Kant's writings accommodate temporal determinacy, since it considers primarily the way in which synthesis arises as an activity. The later Kant's principles of systematicity and purposiveness make their own distinctive contribution, indeed one which more stands in a more natural connection to the terrain of functional relations, rather than only rules in the broader, metaphysical sense in which they appear for much of Kant's writings. This possibility allows traversal of the gulf in meaning between this metaphysical sense and the subjection of events to functional relations which guarantee the occurrence of this or that event based on the occurrence of a predecessor or successor, warranting closer comparisons between the two, and enabling evaluation of the atemporality conjectures to be conducted. Earlier, study of Kant's views on nature's purposiveness revealed that the status of laws in scientific theory is also a product of the mind's organising activity, and that this status ends up being approximate, asymptotic and indeterminate. As a consequence, and as Teufel suggests, these laws can be regarded as revisable. Over time, experiment and discovery prompt revisions to the system of the world rounded out by scientific theory. At any given time, according to Guyer, the theory these laws comprise is incomplete, and destined to be supplemented, modified and superceded in future time.[144] Correlatively, formation of these laws is a process. Thus, Kant discusses what he terms the specific variety, if not the variability, of the laws of nature:

> For it is quite conceivable that, despite all the uniformity of the things of nature according to universal laws, without which we would not have the form of general empirical knowledge at all, the specific variety of the empirical laws of nature, with their effects, might still be so great as to make it impossible for our understanding to discover in nature an intelligible order, to divide its products into genera and species so as to avail ourselves of the principles of explanation and comprehension of one for explaining and interpreting another, and out of material coming to hand in such confusion

(properly speaking only infinitely multiform and ill-adapted to our power of apprehension) to make a consistent context of experience.[145]

Not unlike the way in which Esposito holds the laws governing financial data, such as laws of price movements, to be revisable with changing expectations, even the physical laws which Kant made a consistent focus can too be revised, albeit in a more subtle and complex way. These claims embed temporality into the structure of the theory formation process. Schaper accordingly denounces the temptation to characterise the results of this process as immutable forms, emphasising the potential for discoveries, inventions, errors, paradigm shifts and other disruptions—disruptions which reach out from beyond the territory of science proper, into culture, art, politics and non-discursive modes of thought and understanding—to lead theories to metamorphose.[146] Indeed, Teufel alludes to the possibility of a plethora of changing laws and conditions consisting with the principle of purposiveness—whether adjustments to parameters or paradigm shifts—each being suitably integrated into a coherent theory.[147]

Since functional relations avowedly fall within the purview of the empirical laws Kant's account describes, the impact of the principles of systematicity and purposiveness on the temporality of determinacy can be discerned. This impact is not confined by the limited meaning of temporal determinacy with which these terms appear in the first *Critique*, as was the case in the analysis of apperception, synthesis and the CT; rather, the logical relationship borne to the atemporality conjecture is far more straightforward. Recall its variants, several of which are reproduced below.

Under the mapping $t \rightarrow t - t_0$ solutions of dynamical equations map to solutions of dynamical equations. (Translation Invariance Thesis)
For any law f, and any occurrence of any $t_l \in \{t_i\}$, the truth-value of f does not change with variation in t_l (Truth-Value Constancy Thesis)
For any law f, f is permanent (Permanence Thesis)

Insofar as being a law amounts to performing a specified role in some system of beliefs, constructed according to certain rules, and insofar as such a role changes over time, as new phenomena become explicable or germane or lose salience over time, lawlikeness is impermanent. More formally, if it is a necessary and sufficient condition for a candidate L to be a law that it perform such a role in such a way, a change in this role over time—whether due to discovery or invention of new phenomena which add to, subtract from or otherwise modify the system of beliefs, or due to changes in the rules one applies to construct such a system from what Kant describes as an oceanic multiplicity of possible laws—can change L from a law to a non-law, or from a non-law to a law. This is inconsistent with the Permanence Thesis. What is more, insofar as systematicity opens up a distinction between different times at which different laws are represented as necessary, the translation invariance and truth-value constancy theses become equally hard to maintain. Per Kitcher's framework, a given natural law L could play a specified role R in some system S constructed in accordance with rules $\{M\}$ at a time τ_0 but quite another at time $\tau_1 \neq \tau_0$; at this latter time, some other law might occupy a similar role, or the role itself may have changed in response to observation. If one lets $L = f$, assuming straightforwardly that this law can be characterised as a functional relation, one obtains grounds to reject these two hypotheses. Whilst Kant's analysis is not so specific about whether and how particular occurrences of members of $\{t_i\}$ crop up in laws, if the assignment of truth-values to L over time is in part a function of R, variation in R will induce variation in this assignment, violating the TVC. Analogously, the mapping $t \rightarrow t - t_0$, which maps not only times but by the same token different *roles* to one another $R_t \rightarrow R_{t-t_0}$, and with them laws $L_t \rightarrow L_{t-t_0}$ transforms solutions to *different* solutions of these laws, contra translation-invariance.

Conclusion

This, then, constitutes the Kantian perspective on the temporality of determinacy. The constructive activity of the mind lies at the centre of the synthesis which underpins essential critical doctrines, such as the Analogies of Experience, Schematism and the unity of apperception. This activity is responsible for the subjection of events, the manifold of intuition, to rules which determine them. The Analogies' references to permanence and to causal relations between appearances, whilst the source of an obvious foothold for debates over determinacy, do not categorically settle the question as to whether Kant's work accommodates or refuses temporal determinacy. Looking at the underpinnings of these references in the synthetic facilities of the mind shows how these can be regarded as products of an activity. As a result, the subjection of appearances, events, to rules and laws becomes a process. Then, in Kant's later writings in particular, these themes are developed and extended as he comes to interrogate the contingent or necessary status of empirical laws, and points to a heautonomous, regulative principle which stipulates transcendental conditions for coherence of a scientific theory and its empirical content. This principle posits the unified systematicity or purposiveness of nature and legitimates the representation of laws as necessary within a suitable theory. Kant characterises systematicity and purposiveness as a concept, but nonetheless one which *radiates forth* from the activity of cognition as a process and a quest, rather than a time-independent rule or form.[148] It is a mode of procedure of reflection.[149]

In this sense, Kant's philosophy amounts to a convergence point of two ostensibly clashing lines of thought. It provides the conditions of possibility for the reconciliation of determinism with the denial of the atemporality conjecture and the acceptance of time-dependent laws. This reconciliation consists in Kant's simultaneous affirmation of several propositions. On the one hand, the universality and the necessity of the laws of nature are commitments important enough for them to be cited in several places as a driving force behind Kant's entire project. Experience can contribute only that something is, not that something must be; thus the universality and necessity which we associate with propositions which are not simply mathematical must be accounted for by transcendental conditions. From the first *Critique* to the last Kant refers to laws as transcendentally universal and necessary, whether the subjective, reflective necessity attributed to empirical laws, such as laws of motion, or the objective validity of synthetic a priori judgements. At the same time, Kant respects fully the determinacy conferred on events by these empirical laws, under the guise of causal relations. Every event must have its cause, a principle enshrined in the Analogies of Experience. As noted above, the rules uncovered by Kant in these Analogies are not specific empirical laws and not therefore functional relations. Yet, the relatedness of these empirical relations, or, in Watkins' terms, the lawfulness of these empirical laws, is contributed by the work of understanding.[150] Thus, when Kant claims that there must, for any event, be another preceding event and an accompanying rule this event must follow[151] Kant's account of appearances befits modern accounts of determinism, on the assumption these rules take a functional form, and shorn of the privileged focus on preceding rather than following events:

> A system is deterministic if, and only if, for events $e_1, e_2, \ldots, e_n$ occurring at times $t_1, t_2, \ldots, t_n$ and for any state E_t of this system at any time t, there exists a function relating the $e_1, e_2, \ldots, e_n$, $E_t = f(e_1, t_1, e_2, t_2, \ldots e_n, t_n, t)$.[152] (Def. Determinism)

At the same time that Kant promulgates this quasi-deterministic philosophy and attributes necessity and universality to nature, the subjection of these events to rules is described as a process, in a dual sense—firstly, by virtue of the activity of synthesis; secondly, by virtue of the way in which the mind's activity thinks nature's systematicity and purposiveness. The laws of nature are thus as temporal as is this activity.

In sum, the following three *prima facie* incongruous propositions therefore cohere in Kantian metaphysics:

Laws of nature are transcendentally necessary and universal propositions.
For events occurring at different times, laws of nature exist relating these events to one another, in the form of a function or rule.
Laws of nature, as products of mental activity, are time-dependent.

The journey through post-Kantianism and into the twentieth Century's philosophies of process only superficially amounts to an overturning of the Kantian impulse, and might more properly be regarded as a development of material already contained as a microcosm in his insight. The substitution of dynamic and metamorphosing conditions of possibility in place of ahistorical rules, such as are found in Heidegger's horizon of intelligibility or Nietzsche's genealogy of concepts, is in fact a move in no way alien to transcendental idealism, in which the regulative activity of reason progressively determines events in the world. The historical analysis of scientific theories, and the use of coherentist criteria for the validity of theories and laws, visible in authors from Kuhn to Lakatos to Quine, animates vestiges of Kant's own admission of a panoply of laws, legislated only by the spontaneity of the understanding's authority. This spontaneity belongs to the synthetic activity of the mind, down to which specific acts of synthesis are exemplary of rules, rather than being prescribed by them *ex ante*. Not unlike how Wittgenstein shows that traversing all the instances contained under a rule is no prerequisite of understanding it, one need not believe that Kant holds appearances to be determined to exist or change in advance, by virtue of an atemporal rule. The dissolution of the conditions of possibility of experience and knowledge beyond the periphery of metaphysics and science and into culture, a tendency intensifying with the work of more recent Continental thinkers, is the extremum of one already adumbrated by Kant's critical philosophy, from Hegel's conception of substance as subject, to Cassirer's philosophy of symbolic forms. Such is indeed the result of Schaper's dissection of the Schematism and its consequences—an innovation which implies a mutable structure of nature's patterns, as well as interaction and mixing between the natural world and the formal constraints on human knowledge, hence Kant's identification of nature with the universally lawful nature of experience grounded in the conditions of sensibility and understanding.[153] But unlike these expressions of nomological dynamism, according to which the profile and contours of the world are reshaped over time by thought, through the assertion of nature's uniformity as a necessary principle, Kant's metaphysics is able to preserve the universality and necessity of natural laws in a meaningful sense in the face of this time-dependency, integrating his own species of causal determinism with the temporality of determinacy.

Notes

1. Teufel (2017), p. 115.
2. Kant (1933), A100.

3. Kant (1933), A832/B860.
4. Kant (1933), B196/A157.
5. Kant (1933), A833/B861.
6. Intuition for Kant is that in which objects are given.
7. Kant (1933), B131–132. The guiding principle of the Analogies Kant describes in B220 as resting on the necessary unity of apperception, whilst in A142/B181 he describes the schema of the categories as a pure synthesis effected by the transcendental imagination which works in conformity with the unity of apperception.
8. Kant (1933), A157/B196.
9. In Friedman's case, for instance, a priori laws coupled with sensible content provides the foundation for empirical laws. See e.g. Friedman (1992), p. 174. See also Melnick (1973) and Guyer (2008).
10. Watkins (2019), p. 69.
11. Beck (1981), p. 460.
12. Watkins (2005), p. 189.
13. Kant (1933), A216–218/B263–265.
14. Kant (1933), B220.
15. See e.g. Jacobi (2012), Schopenhauer (2018), Hegel (1958), Heidegger (1997).
16. Schaper (1964), p. 274.
17. Watkins (2005), pp. 185–186.
18. Watkins (2005), p. 213; Kant (1933), A196/B241–242.
19. Kant (2004), p. 29.
20. Kant (1933), A182–185/B225–228.
21. Ginsborg (1997), pp. 39–42.
22. Teufel (2017), pp. 108–112.
23. Teufel (2017), pp. 115–119.
24. See e.g. Tolley (2006) and Warren (2017).
25. Kant (1958), p. 168.
26. Watkins (2017, 2019).
27. Kant (1933), B223.
28. De Pierris and Friedman (2018).
29. A fuller exposition of this framing can be found in Messina (2017).
30. Messina (2017), pp. 133–134.
31. Kant (1933), B224; Guyer (1987), p. 208.
32. Allison (2004), pp. 229–234.
33. Kant (1933), A176–177, A180/B222.

34. Kant (1933), B225.
35. Allison (2004), p. 238.
36. Kant (1933), B226.
37. Kant (1933), A189, B233.
38. Friedman (1992), pp. 170–180.
39. Watkins (2005), pp. 245, 251, 288–289.
40. Kant (2005), pp. 161–162.
41. Kant (2005), p. 175.
42. Guyer and Walker (1990), p. 236.
43. Kant (2004), p. 47.
44. Watkins (2019), p. 2.
45. Kant (1933), A195–196/B240–241.
46. Kant (1933), A549/B577.
47. Beck (1981), p. 453.
48. Kannisto (2017), pp. 495–516.
49. Watkins (2005), p. 257.
50. Guyer (1987), p. 215.
51. Kant (1933), B275–276.
52. Kant (1933), B225.
53. Kant (2005), p. 175.
54. Allison (2004), pp. 239–240.
55. Allison (2004), pp. 245–246.
56. Heidegger (1997), p. 75.
57. Ameriks (2017), p. 37.
58. Guyer and Walker (1990), p. 232. It should be noted that scholars disagree over whether or not Kant's usage of rule and law can be equated. Kannisto, for instance, cites various passages from the *Critique of Pure Reason* such as A91/B124, A113, A126, as well as the *Prolegomena* (312), *Critique of Practical Reason* (20f, 31, 67) and the *Critique of Judgement* (182–184) and *Reflexionen* (5414). For a dissenting view see e.g. Uygar (2020).
59. Kant (1933), A176–177, A180/B222.
60. Kant (1933), B220.
61. Kant (1933), A157/B196.
62. Kant (1933), B134–B135.
63. Kant (2005), p. 261.
64. Kant (1933), A107.
65. Kant (1933), B131–132.

66. Kant (1933), A118, A120.
67. Kant (1933), B131–132.
68. Kant (1933), A119.
69. Pereboom (2019).
70. Strawson (1966), pp. 93–94; Allison (2004), pp. 163–164.
71. Kant (2005), p. 154.
72. Kant (2005), p. 221.
73. Kant (1933), A51/B75, B134.
74. Wolff (1963), p. 96.
75. Wolff (1963), p. 178.
76. Allison (2004), p. 80; Kant (1933), A68/B93.
77. Pippin (1987).
78. Pippin (1987), pp. 454, 463.
79. Watkins (2005), pp. 16, 231, 257, 269. More specifically, Watkins attributes three distinguishing elements to Kant's notion of activity—irreducibility, indeterminacy, and asymmetry. It should also be noted that Watkins repeatedly uses the phrase "temporal determinacy" in a restricted sense: to refer to the application of the categories to appearances which thereby become determinate.
80. Kant (1933), A139/B178.
81. Kant (1933), A142/B181.
82. Schaper (1964), pp. 267–292.
83. Schaper (1964), p. 275.
84. Schaper (1964), pp. 276–282.
85. Land (2018), p. 196.
86. Schaper (1964), pp. 276, 287.
87. De Pierris and Friedman (2018). Also refer to this activity as a construction.
88. Kant further acknowledges the difficulties in interpreting the Schematism in a note from 1797; see Kant (2005), p. 394.
89. Matherne (2014), p. 184.
90. Bird (1962), p. 52; Schaper (1964), p. ss279.
91. See e.g. Matherne (2014), p. 4.
92. Heidegger (1997), pp. 65, 68, 71.
93. Heidegger (1997), pp. 70–72.
94. Heidegger (1997), pp. 121, 123.
95. Kitcher (1982), p. 54.
96. Kitcher (1982), pp. 54–57; Guyer (1980), p. 206.
97. Kitcher (1982), p. 54.

98. Allison, for one, argues that Kant upholds a distinction between the indeterminate forms of intuition and determinate formal intuitions, both of which are types of pure intuitions. See Allison (2004), p. 15.
99. Bader (2017), pp. 12–13.
100. Valaris (2008), p. 15.
101. Kant (1933), A176/B219.
102. Valaris (2008), p. 4.
103. Kant (1933), A31–32/B47–48, A42/B58; Kant (2005), p. 154.
104. Kant (1933), A107.
105. Kant (1933), A143/B183.
106. Kant (2005), p. 227.
107. Kant (2005), p. 289.
108. Heidegger (1997), pp. 80–82.
109. Heidegger (1997), pp. 75–76.
110. Heidegger (1997), p. 76.
111. Heidegger (1997), p. 115.
112. Heidegger (1997), p. 121.
113. Comparably, Hegel, in his Lectures on the History of Philosophy, argues that Kant's claim that time is universal, necessary and a priori does not imply that temporal relations precede the acts which impose them, comparing this process to the process of chewing food. See Hegel (1995), pp. 434–435.
114. Kitcher (1982), p. 54.
115. Petersen (2017), p. 79; Watkins (2005), p. 257; Heidegger (1997), p. 66; Dahlstrom (2010), pp. 396–398.
116. Kant (1933), B158.
117. Kant (1933), A665/B 693, A645/B673.
118. Kant (2007), pp. 15–20, 188.
119. Kant (2007), pp. 19–20.
120. Kant (2007), p. 15.
121. Kant (2004), p. 57.
122. Kitcher (1998), pp. 226–227.
123. Kitcher (1998), p. 222.
124. Guyer and Walker (1990).
125. Kant (1933), A651/B679.
126. Kant (1995), pp. 244–246; Kant (2007), pp. 16, 17; Guyer and Walker (1990), p. 228.
127. Kant (2007), p. 18.

128. Kant (1933), A665–666; Guyer and Walker (1990), p. 246.
129. Kant (2007), p. 16.
130. Kant (2007), pp. 19, 21.
131. Teufel (2017), pp. 116, 127.
132. Kitcher (1998), pp. 226–227.
133. Kant (2007), p. 50.
134. Kant (2007), p. 23.
135. Guyer and Walker (1990), pp. 237, 242.
136. Kitcher (1982), p. 54.
137. Kant (1933), B234.
138. See also for this claim De Pierris and Friedman (2018).
139. Kant (1933), A190/B233.
140. Ginsborg (1997), pp. 59–62.
141. Ginsborg (1997), p. 64.
142. Ginsborg (1997), p. 67.
143. Watkins, for his part, affirms that Kant held temporal determinacy to be a sufficient condition for the work of the category of causality and the requisite activity of the understanding, and conceives of this as an essential aspect of the Second Analogy. See Watkins (2005), p. 257.
144. Guyer and Walker (1990), p. 240.
145. Kant (2007), pp. 20–21.
146. Schaper (1964), p. 289.
147. Teufel (2017), pp. 113, 125.
148. Kant (2007), p. 18.
149. Kant (2007), p. 19.
150. Watkins (2019), p. 39. Importantly, it does not follow from this claim that the particulars of given empirical laws can be derived logically from the principles of the understanding.
151. Kant (1933), A193/B238.
152. Russell (1918), p. 199.
153. Kant (2004, 4:319.

Bibliography

Allison, H. (2004). *Kant's Transcendental Idealism: An Interpretation and Defense.* Yale University Press.

Ameriks, K. (2017). On Universality, Necessity, and Law in General in Kant. In M. Massimi & A. Breitenbach (Eds.), *Kant and the Laws of Nature*. Cambridge University Press.

Bader, R. (2017). Inner Sense and Time. In A. Gomes & A. Stephenson (Eds.), *Kant and the Philosophy of Mind: Perception, Reason and the Self* (pp. 124–137). Oxford University Press.

Beck, L. (1981). Kant on the Uniformity of Nature. *Synthese, 47*(3), 449–464.

Bird, G. (1962). *Kant's Theory of Knowledge*. Humanities Press.

Dahlstrom, D. (2010). The *Critique of Pure Reason* and Continental Philosophy: Heidegger's Interpretation of Transcendental Imagination. In P. Guyer (Ed.), *The Cambridge Companion to Kant's Critique of Pure Reason*. Cambridge University Press.

De Pierris, G., & Friedman, M. (2018). Kant and Hume on Causality. In E. Zalta (Ed.), *The Stanford Encyclopedia of Philosophy*. Retrieved March 7, 2021, from https://plato.stanford.edu/archives/win2018/entries/kant-hume-causality

Friedman, M. (1992). Causal Laws and the Foundations of Natural Science. In P. Guyer (Ed.), *The Cambridge Companion to Kant* (pp. 161–199). Cambridge University Press.

Ginsborg, H. (1997). Lawfulness Without Laws: Kant on the Free Play of Imagination and Understanding. *Philosophical Topics, 25*(1), 37–81.

Guyer, P. (1980). Kant on Apperception and "A Priori" Synthesis. *American Philosophical Quarterly, 17*(3), 205–212.

Guyer, P. (1987). *Kant and the Claims of Knowledge*. Cambridge University Press.

Guyer, P. (2008). *Knowledge, Reason and Taste: Kant's Response to Hume*. Princeton University Press.

Guyer, P., & Walker, R. (1990). Kant's Conception of Empirical Law. *Proceedings of the Aristotelian Society, Supplementary Volumes, 64*, 221–258.

Hegel, G. (1958). *Philosophy of Right* (T. Knox, Trans.). Clarendon Press.

Hegel, G. (1995). *Lectures on the History of Philosophy, Vol. 3: Medieval and Modern Philoosphy* (E. Haldane & F. Simson, Trans.). University of Nebraska Press.

Heidegger, M. (1997). *Kant and the Problem of Metaphysics* (R. Taft, Trans.). Indiana University Press.

Jacobi, F. (2012). *Werke* (Vol. 3). Nabu Press.

Kannisto, T. (2017). Kant on the Necessity of Causal Relations. *Kant-Studien, 108*(*4*), 495–516.

Kant, I. (1933). *Immanuel Kant's Critique of Pure Reason* (N. Smith, Trans.). Macmillan.

Kant, I. (1958). *Critique of Practical Reason* (L. Beck, Trans.). Bobbs-Merrill.
Kant, I. (1995). *Opus Postumum* (E. Förster and M. Rosen, Trans.). Cambridge University Press.
Kant, I. (2004). *Prolegomena to Any Future Metaphysics* (G. Hatfield, Ed.). Cambridge University Press
Kant, I. (2005). *Notes and Fragments* (P. Guyer, Ed. and C. Bowman, P. Guyer, & F. Rauscher, Trans.). Cambridge University Press.
Kant, I. (2007). *Critique of Judgement* (J. Meredith, Trans.). Oxford University Press
Kitcher, P. (1982). Kant on Self-Identity. *The Philosophical Review, 91*(1), 41–72.
Kitcher, P. (1998). Projecting the Order of Nature. In P. Kitcher (Ed.), *Kant's Critique of Pure Reason: Critical Essays* (pp. 219–238). Rowman & Littlefield.
Land, N. (2018). *Crypto-Current, An Introduction to Bitcoin and Philosophy*. Retrieved August 22, 2020, from http://sumrevija.si/wp-content/uploads/2018/11/PDF.pdf
Matherne, S. (2014). Kant and the Art of Schematism. *Kantian Review, 19*(2), 181–205.
Melnick, A. (1973). *Kant's Analogies of Experience*. Chicago University Press.
Messina, J. (2017). Kant's Necessitation Account of Laws and the Nature of Natures. In *Kant and the Laws of Nature* (pp. 131–149). Cambridge University Press.
Pereboom, D. (2019). Kant's Transcendental Arguments. In E. Zalta (Ed.), *The Stanford Encyclopedia of Philosophy*. Retrieved March 21, 2021, from https://plato.stanford.edu/archives/spr2019/entries/kant-transcendental
Petersen, C. (2017). *Kant on Time: Self-affection and the Constitution of Objectivity in Transcendental Philosophy.* Doctoral Thesis. Retrieved February 13, 2021, from http://repository.essex.ac.uk/21752/
Pippin, R. (1987). Kant on the Spontaneity of Mind. *Canadian Journal of Philosophy, 17*(2), 449–476.
Russell, B. (1918). *Mysticism and Logic and Other Essays*. Longmans, Green and Co.
Schaper, E. (1964). Kant's Schematism Reconsidered. *The Review of Metaphysics, 18*(2), 267–292.
Schopenhauer, A. (2018). *Sämtliche Werke*. Forgotten Books.
Strawson, P. (1966). *The Bounds of Sense: An Essay on Kant's Critique of Pure Reason*. Methuen.
Teufel, T. (2017). Kant's Transcendental Principle of Purposiveness and the "Maxim of the Lawfulness of Empirical Laws". In M. Massimi &

A. Breitenbach (Eds.), *Kant and the Laws of Nature*. Cambridge University Press.

Tolley, C. (2006). Kant and the Nature of Logical Laws. *Philosophical Topics, 34*, 371–407.

Uygar, A. (2020). *Kant on Laws*. Notre Dame Philosophical Reviews. Retrieved April 19, 2021, from https://ndpr.nd.edu/reviews/kant-on-laws/

Valaris, M. (2008). Inner Sense, Self-Affection and Temporal Consciousness. *Philosophers' Imprint, 8*(4), 1–18.

Warren, D. (2017). Kant on Mathematical Force Laws. In *Kant and the Laws of Nature* (pp. 171–192). Cambridge University Press.

Watkins, E. (2005). *Kant and the Metaphysics of Causality*. Cambridge University Press.

Watkins, E. (2017). Kant's Necessitation Account of Laws and the Nature of Natures. In *Kant and the Laws of Nature* (pp. 11–29). Cambridge University Press.

Watkins, E. (2019). *Kant on Laws*. Cambridge University Press.

Wolff, R. (1963). *Kant's Theory of Mental Activity: A Commentary on the Transcendental Analytic of the Critique of Pure Reason*. Harvard University Press.

6

Conclusion

The atemporal philosophy of laws has an intuitive starting point: the world is a system of events, and such events are described by laws. Regardless of their other characteristics, these laws take the form of mathematical functions, or relations of sets of events. Subjection to these laws determines an event to occur here or there in time, and to be determinate is to conform to a function. The uniformity and consistency of nature, taken as a necessary condition for determinism, for the intelligibility of the world, or for the validity of scientific enquiry itself, precludes variation in these laws, whether or not they are ultimately rejected or overturned. Appearances of chaos or contingency harbour no implications for them; assuming they exist or can be found, they do not change, no matter how transformative may be the eonic shifts in the behaviour of the world, whether between cosmological phases or from universe to universe. Their atemporality is constituted in this lack of variation, permanence, and transposition through seemingly distinct epochs.

Constraints are then taken to apply as to their corresponding functional form. The principles of consistency and uniformity, as well as faithfulness to determinism, require time to disappear from formal descriptions of laws, a necessary condition for the integrity of causal or functional relationships. The temporality of the world is constituted in the indexation of events by labels, whether or not one takes this indexation as

C. Husbands, *The Temporality of Determinacy*,
https://doi.org/10.1007/978-3-030-86530-6_6

having to do with the flow of time as an independent object, without regard, as Newton claimed, to anything outside of itself.[1] In turn, the eternality of these functions implies they ought to be free of such indexation. Time thus crops up only superficially in the argument of functions, representing the failure of a system to be causally closed, appropriately isolated from perturbations from the outside, or an incomplete understanding of the relevant dynamics. Epistemic limitations alone prevent the researchers of today from grasping the fundamental relations to which these functions will ultimately be reduced, and such appearance reflects no underlying dependency on time on the part of the true variables. The form of laws and their truth values must be preserved over different times, whether such differences lie on the quantum or cosmological scale; a translation through time which generated different such laws would attest to the fact that they are formed improperly, and fall short of the unlimited scope becoming of a permanent rule. These sentences express the view that the laws of nature must be absolutely independent of time, denying the temporality of determinacy.

This work argues against this philosophy. Firstly, it decomposes the atemporality conjecture into four sub-doctrines, each of which gives it a more precise sense. Secondly, it attempts to trial these sub-doctrines against a number of theories in the physical and social sciences, identifying respects in which they are inconsistent, and in which there is a valid sense in which these theories incorporate time-dependent descriptions. Thirdly, it attempts to undermine the link, posited by the philosophy of atemporality, between the principles of uniformity and consistency on the one hand and the time-independence of the subject laws on the other, by showing that these can cohere in a canonical philosophical system, such as transcendental idealism.

This argument proceeds in the following stages. The second chapter supplies a number of the definitions which are central to the argument of the book. It draws upon the work of Russell to present functions relating events as the primary language in which scientific laws are expressed, and defines determinacy as subjection to these functions. This leads to the following definitions of determinism and determinacy, which are employed throughout the rest of the book (unless stated otherwise).

A system is deterministic if, and only if, for events $e_1, e_2, \ldots, e_n$ occurring at times $t_1, t_2, \ldots, t_n$ and for any state E_t of this system at any time t, there exists a function relating the $e_1, e_2, \ldots, e_n$, $E_t = f(e_1, t_1, e_2, t_2, \ldots e_n, t_n, t)$.[2] (Def. Determinism)

An event has determinacy iff it is subject to a functional relation f. (Def. Determinacy)

It then considers the parallels between Russell's work and the work of other scholars in twentieth century philosophy of science, such as Lindsay and Margenau and A. Ushenko, but which can also be identified in, for example, Grünbaum's response to these authors.[3] These provide the clearest recent concretization of the antipathy towards time-dependent laws, and enable several interrogable theses which express the atemporality conjecture to be articulated.

Under the mapping $t \rightarrow t - t_0$ solutions of dynamical equations map to solutions of dynamical equations. (Translation Invariance Thesis)
For any law f, and any occurrence of any $t_l \in \{t_i\}$, the truth-value of f does not change with variation in t_l (Truth-Value Constancy Thesis)
For any law f, f is permanent (Permanence Thesis)
For no law f does the time-variable t occur in f where $dg'' + bg' + cg + d = f$ and $g' = \frac{\partial g}{\partial t}$, $g'' = \frac{\partial^2 g}{\partial t^2}$. (Time-Variable Thesis)

The subsequent two chapters consider respectively the non-physical and the physical sciences, and provide an extended analysis of developments in these fields which (i) refute these sub-doctrines and (ii) provide a sense in which the systems concerned might justifiably regarded as time-dependent. Both (i) and (ii), this book argues, motivate the rejection of the atemporality conjecture, and the absolute time-independence assigned to laws.

The analysis of the non-physical sciences refers to four main theories, or programs of research. First of all, it notes how recent research in the foundations of finance and economics focuses on the metaphysics of time implicit in these disciplines, using these assumptions to signal limitations of these theories and open up alternatives. More specifically, the chapter considers the work of Elena Esposito. Esposito's theses of endogeneity and revisability are the key premises in the following argument:

1. Relations of time are internal to systems
2. Expectations about prices comprise financial systems
3. The future is bounded by the present future and future present
4. The present future consists of expectations about prices and price movements
5. The future is the revisability of the future present and the present future[4]

This argument establishes two senses in which the participatory systems described by Esposito might be regarded as time-dependent. First of all, the determination of events in this system is a process, as it admits of an order. Second of all, it proposes a distinction between the order in which events occur and the order in which they are determined. Whilst events occur in an order which can be indexed by ordinary time variables $\{t_i\}$, these are distinct from the order in which distinct functions come to apply to them. The two are not related monotonically.

Following this section, another prominent tool in the social sciences is considered, namely time series analysis. Here, time is an explicit part of the functions used to model complex systems with many dependencies. These functions are often probabilistic, and can exhibit characteristics such as autocorrelation, heteroskedasticity, and non-stationarity, which present serious challenges to the properties espoused by the atemporality conjecture. Time-dependency is therefore embedded in the essence of the descriptions and techniques wielded by the discipline. Thirdly in this chapter, the contributions of Massumi to international relations and military theory are considered. Massumi identifies two aspects of the operative logic of twenty-first century conflicts which exhibit temporal determinacy. First, pre-emptive attacks are based on iterated modal propositions about the state of an enemy actor—propositions about possibilities which could have been the case, or about what always will have been possible. Plausible logical formulations of these propositions cannot be reconciled with at least two of the atemporality sub-doctrines, namely truth-value constancy and translation-invariance. Second, Massumi develops a concept of future causation, visible in a simple model of an exchange between two threat actors: by virtue of a retaliatory strike, a previous possibility is actualized, and an undecideable threat becomes a germinal attack. This concept implies a time lag over which earlier events become determined, and therefore a similar distinction to Esposito's—between the order of occurrence of the events in a sequence and the order in which they are determined. A final example of this distinction—a prevalent one in the non-physical sciences—is then expounded, drawn from cryptography. Here, too, the order of events, each event corresponding to a financial transaction, is impermanent and determined through a process. It changes between different protocols, different distributions of computing power, or participants' changing volitions, pointing to a time in which these regimes change, as well as a time in which events occur.

Examples from the physical sciences are the subject of the fourth chapter. It notes the growing interest in theories according to which laws evolve between different cosmological epochs, either due to the variation in fundamental physical parameters such as the fine-structure constant, or due to mechanisms which lead to the production of universes through singularities, each with laws different to its progenitor due to small, random perturbations. Certainly, to uphold any aspect of the atemporal philosophy in the context of these theories would require an inventive reinterpretation: the transition from universe to universe, with previous laws rendered impermanent, relativized to particular cosmological epochs, and changing truth value and form, clashes with the spirit of the conjecture as well as the letter of its sub-doctrines. Secondly, this chapter attends to an established area of physics in which probabilistic concepts surface regularly, and which often attracts the attention of philosophers of time, namely statistical mechanics. It introduces a framework of dynamical systems theory, which characterises a system of interest as a triple $[X, \mu, T_t]$, subject to the following definitions:

$\{T_t \mid t \text{ is an instant of time}\}$ is a family of automorphisms, or transformations of X onto X

T_t is such that $T_{t_1+t_2} = T_{t_1}\left(T_{t_2}\right)$ for all $x \in X$.[5]

It then considers an important class of systems studied in this area, namely ergodic systems. Using the research of Charlotte Werndl, Roman Frigg and others, it is shown that a type of ergodic system displays probabilistic irrelevance, in the following sense: Pairs of events represented by the two variables *X*, *Y* occurring respectively at a preceding time t_{-n} $(n > 0)$ and later zeroth time t_0 are such that:

$$\lim_{n\to\infty} p\left(X^{t_0} \mid Y^{t_{-n}}\right) - p\left(X^{t_0}\right) = 0 \left(\mathrm{PI2}\right)$$

In essence, this means that the probabilities of the latter event happening given that the former event happens differ negligibly from the *overall* probability of the latter event happening. This underscores the conclusion of the analysis of time series analysis, in that subjection to laws need not mean determination of all events at all times. Because these laws incorporate statistical, probabilistic relationships between events, they become determinate within finite time intervals, or, in the extreme case, as and when they occur.

After this, the survey of the physical sciences considers another field in which unresolved metaphysical antinomies are seemingly abundant. The relationship between time and determinacy in the context of quantum mechanics has been taken up by some of its most distinguished contributors. Recently, philosophers have gravitated towards the quantum aspect of a notion of retrocausality. The section explores Huw Price's time-symmetry argument for retrocausality (TSAR), which attempts to show that the conjunction of several plausible theses entails retrocausality, thus:

Premise (1) Realism
Premise (2) Time-symmetry
Premise (3) Discreteness
Conclusion (C) Retrocausality

It then considers a generalized version of Price's argument, under which certain assumptions are relaxed and others refined, put forward by Leifer and Pusey in the last decade. These authors claim that the following seemingly plausible no-retrocausality thesis is violated for a class of experimental set-ups.

$p(a, b, \lambda | x, y) = p(b | a, \lambda, x, y) p(\lambda | a, x) p(a | x)$ (No Retrocausality Thesis NRT)

Consequently, the probabilities of output results occurring earlier in time than certain variables are not independent of these variables, which is to say that $p(\lambda | a, x, y) = p(\lambda | a, x)$ does *not* hold (in spite of the fact that $t_y > t_\lambda$). These considerations, it is argued, imply a distinction between the order in which the measurement outcomes or events occur and the order in which they are determined, in a manner which parallels the attack on the atemporality conjecture presented in previous chapters. Time-dependency is once again constituted in the fact that the determinacy to which events are subject comes about in an order, and could be considered a process. The section concludes by considering other aspects of retrocausality in the quantum domain, such as the transactional interpretation of quantum mechanics, before closing the chapter by examining whether time-dependency in the sense outlined is consistent with relativity theory and Noether's theorem.

The penultimate chapter turns to the philosophical implications of these studies. It attempts to demonstrate that precedent exists for the temporality of determinacy within modern philosophy. This provides a counter-example to two claims: firstly, that the temporality of determinacy is inconsistent with a deterministic worldview; secondly, that the intuitive principles of uniformity and consistency (espoused by the likes of Lindsay and Margenau) are sufficient conditions for the atemporality conjecture. It considers two main aspects of Kant's thought. The *Critique of Pure Reason* examines in detail the conditions for the time-determination of appearances and the subjection of appearances to rules, in the Analogies, the doctrine of the unity of apperception, and in the Schematism. It is shown that these functions are, for Kant, ultimately rooted in the processual, synthetic activity of the mind. A constructionalist interpretation of this activity, following the analysis of Eva Schaper, is defined as follows:

> Given two states S_1 and S_2 there is a further state S_3 such that S_3 would either (i) not exist or (ii) be different, had one or both of S_1 or S_2 either (i) not existed or (ii) been different, and S_3 is an activity.[6] (Constructionalism Thesis)

From this perspective, synthetic activity enables temporal determinacy in the dual sense that S_3 is a process, and itself depends on some other parameter τ which varies throughout this process. The temporality of determinacy thus enters into Kantian discourse once this activity is established as a process which occurs at one time rather than another.

Secondly, the chapter considers Kant's later work and, in particular, the implications of the *Critique of Judgement.* Here, Kant develops his principles of systematicity and purposiveness, which consider the necessary conditions for laws of nature, and how the contingency of empirical laws—to be distinguished from the pure *a priori* laws of the understanding—with the seeming necessity they impart to theories containing them. The chapter considers various scholars' conceptions of these principles, from Kitcher, Messina and Teufel to Guyer, Walker and Ginsborg. It presents the following set of tenets which characterise the principles of purposiveness and systematicity.

A given law L constitutes a law of nature if and only if L performs a specified role R in some system of beliefs S constructed in accordance with a set of rules $\{M\}$

L is represented as necessary if and only if it plays this role

L is contingent

The role R in S in accordance with rules $\{M\}$ is performed in a way which is asymptotic, indeterminate, and approximate.

1. is a transcendental principle.[7]

The relation to temporal determinacy is secured by the realization that these functions are enacted by the understanding in a procedure. The systematicity of the laws of nature is therefore an activity no less than is the synthesis of the manifold in Kant's earlier critical works. The understanding conforms to the principle of purposiveness through a process in which it works actively on the belief system (S). Laws must be regarded as necessary, but the role they play which qualifies them as such involves their perpetual formation, variation and reformation, all driven by the activity of the mind. Kant thus upholds fully the determinacy conferred on events by these empirical laws, under the guise of the causal relations described in the Analogies. He upholds the universality and the necessity of the laws of nature, and presents the principle of purposiveness as a transcendental rule staving off the risk of chaos and degeneracy. And, in this sense, he enables the reconciliation of the temporality of laws with determinism and the uniformity, consistency, and rationality of nature bound up with it: in his later thought the latter become, if anything, *grounds* for the former rather than objections to them.

Origins of Temporality and Determinacy

Supposing this argument to be correct might lead one to question the provenance of the philosophy of atemporality, and why the alternative put forward in this book at first clashes with one's intuitions and feels foreign to metaphysics and science. From a diachronic standpoint, this atemporality stretches forward from as far back as Ancient thinking on determinism, and through almost every epoch since. Examples of this are easy to find, and offer some insight into the origins of its grip on our thinking.

A natural place begin is with the Roman philosopher Cicero. His writings, such as *De Fato*, preserve Stoic and sceptic perspectives on causality, time and logic, and respond to a question familiar for these thinkers: can anything occur which was not already determined to occur in time? Other metaphysicians such as Chrysippus, held logical bivalence to require preceding and therefore eternal causes.[8] That every proposition need be either true or false required jettisoning fortuitous causes, or what was known as spontaneity in the relation of cause to effect. Cicero, however, aims to separate these two commitments. He contends instead that the deterministic worldview, understood as nature's conformity to rules, is compatible with fortuity, spontaneity and time-dependency—precisely the features of the causal relation rejected by his Stoic antagonist. In so doing, he finds that causal relations need not be eternal.

The Stoics' views on time and causality were not the only ones Cicero accuses of conflating logical and metaphysical conclusions. In Epicureanism, too, he locates a false inference from bivalence to time-independent causation. Given the tenet that every proposition must be either true or false, it does not follow that all events are necessarily brought about and determined by fate. By contrast, Epicurus, Cicero holds, infers from bivalence the agency of an eternal chain of necessary natural causation which stands outside of the ravages of time.[9] In its place he puts up an alternative: a compatibilist doctrine which holds, not between freedom and determinism, but between determinism and the time-dependence of the laws of nature, and of the causal powers they exert.

Certainly, Cicero was not the only author to explore such lines of thought. A similar perspective can be found in Plutarch's *De Stoicorum Repugnantiis*.[10] Plutarch's remarks oppose two positions. The first is that of a foil who violates nature by proposing a theory of the uncaused, that is, by pointing to ostensibly random behaviour, such as that of a die, as if to identify a system which flouts natural laws. The other position upholds a deterministic principle in the face of this behaviour, namely uniformity: a necessary condition for a difference in the outcome of a putatively causal interaction is a change in the initial conditions. Conversely, like causes produce like effects. This principle, referred to variously as the Specified Causal Principle,[11] the causality principle,[12] or one which links same cause to same effect,[13] Plutarch's passage holds to be consistent with spontaneous processes which belie underlying regularities, just as does the die cast.[14] Such processes indicate different *initial* conditions, whether those of the environment external to the events being compared, or the latent state of the things themselves. In this extract from Plutarch, unlike in Cicero's presentation, no appeal to bivalence is needed, but only the absurdity of uncaused motion. Consequently, he holds that causal principles do not preclude motion which is not pre-determined, or which is time-dependent.

Moving through the Middle Ages, authors working in the tradition of theological determinism wrestled with the implications of determinism in the context of predestination, eternity, salvation and damnation. Spinoza provides a characteristic example, endorsing a necessitarianism with clear implications for the relationship between eternity and causality, and time and determinacy.[15] Events, as modifications of substance, not only follow necessarily from their causes, but follow from causes which are themselves necessary. Precisely, Spinoza holds that for all substances x, if at time t the substance x has the property of being F, then x necessarily has F at t.[16] Provided we are willing to gloss a substance's having a property as an event—a permissible simplification in the context of this appropriation of Spinoza's metaphysics, in which events are modes or affections of substance—then all events are necessary; no alternative sequences are possible. Thus, Spinoza claims, there is nothing contingent in nature, and the divine nature determines things to exist and to exert whatever powers by necessity.[17] And as respects the relation between this kind of necessity and temporality, Spinoza's position ensures that necessity is related intimately to eternity. Everything which follows from God's power and nature flows from eternity to eternity with the same necessity as geometric theorems.[18] Neither time nor duration, in Spinoza's terms, have any ability to explain existence.[19] Since all events are necessary, and necessary truths are eternal, his position allows no room for temporal variation in the truth values of propositions about these events (nor, by the same token, variation in the truth values of physical laws). That a given event occurs, when and how it occurs—these determinations are *independent of time*.

Many more historical thinkers can be brought into this debate over time, determinism and their logical relationship. They are past epochs' equivalents of the combatants cited in modern physical and non-physical science and philosophy. Of course, there are differences between these epochs and the present day which limit the parallels which can be drawn legitimately. It was not until after both the ancient and scholastic periods that philosophical debates about laws were philosophical debates about *physical* laws, imbued with the thoroughgoing focus on the nature of the world, rather than merely the nature of man, which characterises them today.[20] Only with Galileo's work did the modern view of laws and rules as functional relations which connect physical quantities to one another[21] come about. Thus, the ancients' discussions related less to the determinacy of events subject to laws over time, and more to the truth or falsity of propositions over time. They refer less to time and determinacy and more to eternity and causality. Yet despite differences in the conception of laws from epoch to epoch, what they have in common is that each contributor affirms or denies the temporality of the rules or laws in virtue of which events making up the system of the world—or the propositions about these events—are as they are.[22] As a result, identifying their counterparts among today's philosophers of science is a straightforward exercise. Lindsay and Margenau require nature to be time-independent in order to be consistent and uniform and to avoid the chaos of a non-causal world subject to the irregular whims of unknown powers.[23] This in turn they connect to their thesis, criticised throughout this book, that the differential equations used by scientists to describe the world should not contain functions of time. Russell, in "On the Nature of Causality," invokes a similar principle, discussed in the second chapter, according to which functional relations between events remain true with respect to variation in *t*. These and many others are the heirs of the metaphysics of atemporality expressed in thousands of years of philosophical exchanges.

Based on these examples, it is plausible that this vision of atemporality is not specific to contemporary determinism, but recapitulates a tradition. Epicurus' thought places the stream of causation, the linkages between events brought about by some continuous action, outside of time—a necessary condition, allegedly, for avoiding a theory of the uncaused. Spinoza ascribes an eternal necessity to all worldly events, by virtue of an eternal power of divine action.[24] Henry of Huntingdon cites King Canute's reference to eternal laws as evidence of his piety: the divine will is expressed not only in its might and ubiquity, but in its escape from the decay and variation of worldly temporality. Whether these eternal laws intervene in the world in order to direct and guide phenomena, or whether the phenomena somehow mimic or express the eternal laws as they unfold in time, the laws themselves are no more subject to time than God is finite in extension or limited in power. Whilst the contexts of Ancient philosophy and rationalism predate the shift towards conceiving of laws as functions, or its full reception, the way in which functions fix the entire evolution of the system of the world, determining past and future events from without, mimics in some ways the relation of the powers of nature and will of man to God: what is determined is temporal and worldly, whereas what determines is neither and acts *sine loco*, generating familiar problems of interaction and questions as to how the two sets of characteristics can be reconciled.

In a sense, therefore, the abstraction of the laws of nature from the system they describe displays marks of the Ancient thought of eternal causation, and might seem to be a natural axiom from the point of view of tradition. Taking laws to describe events allows them to be positioned outside of the world, becoming as temporal or atemporal as the functional relations in terms of which they are cast, in spite of the possibility that relations change over time. Acting from beyond appearances, yet knowable by observation, functions enforce events' conformity to rule from moment to moment. In spite of its occasionalist aspects, only action at such a remove from the world is taken by the philosophy of atemporality to preserve the sanctity of determinism and causality, the consistency and uniformity of nature, and the intuition of necessity and universality with respect to laws. This position, attested to by the likes of Lindsay and Margenau, therefore goes hand-in-hand with a conception of laws by virtue of which they are abstracted from physical and social systems; time indexes events, and either evolves or is evolved by the world, but has no purchase on the laws of nature. Dynamics characterize the evolution of systems, but dynamics themselves are no system, nor do they evolve. This is the result of a perspective according to which laws are an inert mechanism, a static set of relations between events, and separated by an impermeable boundary from the world.

Reasons of philosophical principle aside, recent science and philosophy provides us with many reasons to be sceptical about this vision. In respect of the particular sub-doctrines of the atemporality conjecture, this book contends that a suite of examples from recent and less recent science and philosophy coheres poorly with them, or refutes them outright. From statistical mechanics to the foundations of economics, from methods in time-series analysis to research in cryptography, exponentiating advances in a number of fields reveal it to be an anachronism. Instead, drawing upon examples not only from these fields but also from philosophers central to traditional debates about natural or causal laws such as Kant, an alternative position can be argued for: laws of nature need not be time-independent and temporal dependencies need not be expunged for the equations they inhabit to be well-formed, and the increasing centrality of time-dependent descriptions in the sciences does not delegitimate them. More generally, however, the philosophy of atemporality can be understood not only in terms of its particular sub-doctrines such as the TVC or TVT theses, but in relation to the metaphysical context in which it arises. As more schools in the philosophy of science study the conditions of possibility for theory formation, and discover in these conditions not only data and logical relationships but heuristics, temporary auxiliary hypotheses, dynamics of holistic confirmation and refutation, instances of pragmatism, and the deployment of quasi-aesthetic criteria—such as the *simplicity* to which Einstein frequently alluded as a criterion for foundational principles—fewer give credence to a hard boundary between theory and evidence, and therefore between laws and nature. Fewer still would put faith in the permanence of a theory of everything which might fix the dynamics of the systems it describes from outside, as if from the vantage point of a realm of forms. Yet it is the tradition which affirms exactly these residues of theism, according to which theories are exogenous to the phenomena they describe, in which the philosophy of atemporality operates and of which, in some respects, it appears to be an inheritor.

Today, there is no reason to believe that changes in the course of nature need be effected by eternal causes, outside of time, in order to avoid indeterminacy. Nor must one endorse a transcendent power which determines all past and future events in one and the same moment, or rely on absolute time as an *explanans*—especially as the borders between theory and nature become increasingly permeable, and as a thoroughgoing integration is sought between observer and experiment, both within science and in the philosophy of science. A perspective of this sort might be aligned with how some philosophers have come to regard the act of rule-following as expressions of a dynamic form of life as opposed to instances pre- and post-determined by a meta-linguistic absolute. In a formulation from the *Tractatus*, Wittgenstein claims that, if all objects are given, then at the very same time all possible states of affairs must be given, as if to imitate the Laplacian hypothesis.[25] But in his later philosophy, the notion of an act of meaning which traverses in advance all the steps involved in its application, anticipating and determining eternally every instance subject to it in past and future time, is cast aside as a metaphysical crux. Such would be the consequence for atemporal laws of nature, just as for eternal causation, of the same affirmation of temporal determinacy within metaphysics as is already discernible in areas of the sciences.

Notes

1. Newton (1999).
2. Russell (1918), p. 199.
3. Grünbaum (1953), p. 118.
4. Husbands (2020), p. 117.
5. Frigg et al. (2020).
6. Kitcher (1982), p. 54.
7. Kitcher (1998), pp. 226–227.
8. Cicero (1853), p. 273.
9. Cicero (1989), IX 19.
10. Plutarch (1952), 1045c.
11. Bobzien (1998), p. 42.
12. Lindsay and Margenau (1963), p. 522.
13. Russell (1918), p. 91.
14. Bobzien (1998), p. 40.
15. Zilsel accords Spinoza a significant role in the development of the notion of laws of nature as the first to combine metaphysical determinism with a concept of natural law which coheres with its contemporary meaning. See Zilsel (2003), p. 117.
16. This staging is owed to the Spinoza scholar Koistinen (2003), p. 285.
17. Spinoza (1988), p. 433.
18. Spinoza (1988), p. 426.
19. Spinoza (1988), p. 408.
20. For an extended account of this difference see Zilsel's account, "The Genesis of the Concept of Physical Law." Zilsel (2003).
21. Zilsel (2003), p. 111.
22. As Susanne Bobzien shows in her illuminating study of Stoic thought, despite the undeniable contextual and conceptual differences between determinism in its ancient and modern forms, there are scant grounds to deny wholesale that Chrysippus' determinism was a causal determinism, enhancing the comparability of these forms.
23. Lindsay and Margenau (1963), p. 522.
24. Spinoza (1988), p. 442.
25. Wittgenstein (2012), p. 30.

Bibliography

Bobzien, S. (1998). *Determinism and Freedom in Stoic Philosophy*. Oxford University Press.

Cicero, M. (1853). On Fate. In C. Yonge (Trans.), *The Treatises of M. T. Cicero* (pp. 264–282). Henry G. Bohn.

Cicero, M. (1989). On Fate. In H. Rackham (Trans.), *Rhetorical Treatises: Book III, De Fato v. 4: De Oratore*. Loeb.

Frigg, R., Berkovitz, J., & Kronz, F. (2020). The Ergodic Hierarchy. In E. Zalta (Ed.), *The Stanford Encyclopedia of Philosophy*. Retrieved November 1, 2020, from https://plato.stanford.edu/archives/fall2020/entries/ergodic-hierarchy/

Grünbaum, A. (1953). Some Remarks on Professor Ushenko's Interpretation of Causal Law. *The Journal of Philosophy, 50*(4), 115–120.

Husbands, C. (2020). Esposito's Temporality of Finance: Endogeneity and Revisability in Derivative Transactions. *Finance and Society, 6*(2), 114–129.

Kitcher, P. (1982). Kant on Self-Identity. *The Philosophical Review, 91*(1), 41–72.

Kitcher, P. (1998). Projecting the Order of Nature. In P. Kitcher (Ed.), *Kant's Critique of Pure Reason: Critical Essays* (pp. 219–238). Rowman & Littlefield.

Koistinen, O. (2003). Spinoza's Proof of Necessitarianism. *Philosophy and Phenomenological Research, 67*(2), 283–310.

Lindsay, R., & Margenau, H. (1963). *Foundations of Physics*. Dover Publications.

Newton, I. (1999). *The Principia: Mathematical Principles of Natural Philosophy* (I. Cohen & A. Whiteman, Trans.). University of California Press.

Plutarch. (1952). *De Stoicorum Repugnantiis* (M. Pohlenz & R. Westman, Eds.), VI, 2.

Russell, B. (1918). *Mysticism and Logic and Other Essays*. Longmans, Green and Co.

Spinoza, B. (1988). Ethics. In E. Curley (Trans.), *The Collected Works of Spinoza: Volume I*. Princeton University Press.

Wittgenstein, L. (2012). *Tractatus Logico-Philosophicus* (C. Ogden, Trans.). Dover Publications.

Zilsel, E. (2003). The Genesis of the Concept of Physical Law. In *The Social Origins of Modern Science* (pp. 96–122). Kluwer Academic Publishers.

Bibliography

Abbott, A. (2001). *Time Matters: On Theory and Method*. University of Chicago Press.

Abbott, A. (2016). *Processual Sociology*. University of Chicago Press.

Abdussattar, & Vishwakarma, R. (1997). Some Robertson–Walker Models with Variable G and Λ. *Australian Journal of Physics, 50*, 893–901.

Adlam, E. (2018). Spooky Action at a Temporal Distance. *Entropy, 20*, 41.

Aharonov, Y., & Vaidman, L. (2007). The Two-State Vector Formalism of Quantum Mechanics: An Updated Review. Retrieved December 13, 2020, from https://arxiv.org/pdf/quant-ph/0105101.pdf

Albert, D. (1992). *Quantum Mechanics and Experience*. Harvard University Press.

Albert, D. (2000). *Time and Chance*. Harvard University Press.

Alborn, T. (1989). Peirce's Evolutionary Logic: Continuity, Indeterminacy, and the Natural Order. *Transactions of the Charles S. Peirce Society, 25*(1), 1–28.

Allison, H. (2004). *Kant's Transcendental Idealism: An Interpretation and Defense*. Yale University Press.

Ameriks, K. (2017). On Universality, Necessity, and Law in General in Kant. In M. Massimi & A. Breitenbach (Eds.), *Kant and the Laws of Nature*. Cambridge University Press.

Anderson, D., & Zalta, E. (2004). Frege, Boolos, and Logical Objects. *Journal of Philosophical Logic, 33*(1), 1–26.

C. Husbands, *The Temporality of Determinacy*,
https://doi.org/10.1007/978-3-030-86530-6

Augustine of Hippo. (1979). *St. Augustine's Confessions* (W. Watts, Trans.). Heinemann.

Avanessian, A., & Malik, S. (n.d.). The Time-Complex. *DIS Magazine*. Retrieved September 26, 2019, from http://dismagazine.com/discussion/81924/the-time-complex-postcontemporary/

Ayache, E. (2010). *The Blank Swan: The End of Probability*. John Wiley & Sons.

Ayache, E. (2015). *The Medium of Contingency: An Inverse View of the Market*. Palgrave Macmillan.

Ayache, E. (2016). Response to Johnson: A Random Sample Versus the Radical Event. *Finance and Society, 2*(2), 205–216.

Back, A. (2002). *Hashcash—A Denial of Service Counter-Measure*. http://www.hashcash.org/papers/hashcash.pdf

Bader, R. (2017). Inner Sense and Time. In A. Gomes & A. Stephenson (Eds.), *Kant and the Philosophy of Mind: Perception, Reason and the Self* (pp. 124–137). Oxford University Press.

Barabash, A. (2000). Possible Evidence for a Time Variation of the Weak-Interaction Constant from Double-Beta-Decay Experiments. *Physics of Atomic Nuclei., 63*(7), 1210–1213.

Barabash, A. (2003). Possible Evidence of Time Variation of Weak Interaction Constant from Double Beta Decay Experiments. *Astrophysics and Space Science, 284*(4), 607–612.

Barbour, J. (1989). *Absolute or Relative Motion? Vol. 1: The Discovery of Dynamics*. Cambridge University Press.

Barbour, J. (1999). *The End of Time: The Next Revolution in Physics*. Oxford University Press.

Barbour, J. (2001). On General Covariance and Best Matching. In C. Callender & N. Huggett (Eds.), *Physics Meets Philosophy at the Planck Scale* (pp. 199–212). Cambridge University Press.

Barrett, J. (1999). *The Quantum Mechanics of Minds and Worlds*. Oxford University Press.

Barrow, J. (1996). Time-varying *G*. *Monthly Notices of the Royal Astronomical Society., 282*(4), 1397–1406.

Barrow, J., Magueijo, J., & Sandvik, H. (2002). A Cosmological Tale of Two Varying Constants. *Physics Letters B, 541*(3–4), 201–210. https://arxiv.org/abs/astro-ph/0204357

Barrow, J., & Tipler, F. (1986). *The Anthropic Cosmological Principle*. Oxford University Press.

Beck, L. (1981). Kant on the Uniformity of Nature. *Synthese, 47*(3), 449–464.

Bell, J. (1987). *Speakable and Unspeakable in Quantum Mechanics: Collected Papers on Quantum Philosophy*. Cambridge University Press.

Berkovitz, J., Frigg, R., & Kronz, F. (2006). The Ergodic Hierarchy, Randomness and Hamiltonian Chaos. *Studies in History and Philosophy of Modern Physics, 37*, 661–691.

Bird, G. (1962). *Kant's Theory of Knowledge*. Humanities Press.

Blundell, K., & Blundell, S. (2010). *Concepts in Thermal Physics*. Oxford University Press.

Bobzien, S. (1998). *Determinism and Freedom in Stoic Philosophy*. Oxford University Press.

Bohr, N. (1950). On the Notions of Causality and Complementarity. *Science, 111*(2873), 51–54.

Bokulich, A., & Bokulich, P. (2020). Bohr's Correspondence Principle. In E. Zalta (Ed.), *The Stanford Encyclopedia of Philosophy*. https://plato.stanford.edu/archives/fall2020/entries/bohr-correspondence

Bonneau, J., et al. (2015). *SoK: Research Perspectives and Challenges for Bitcoin Cryptocurrencies*. IEEE Symposium on Security and Privacy. http://www.jbonneau.com/doc/BMCNKF15-IEEESP-bitcoin.pdf

Brading, K., & Brown, H. (2003). Symmetries and Noether's theorems. In K. Brading & E. Castellani (Eds.), *Symmetries in Physics: Philosophical Reflections* (pp. 89–109). Cambridge University Press.

Brassier, R., et al. (2017). Pricing Time: Outline and Discussion on Suhail Malik's "The Ontology of Finance". *Identities: Journal of Politics, Gender and Culture, 14*(1), 84–128.

Brock, W. (2001). *Growth Theory, Nonlinear Dynamics and Economic Modelling* (W. Dechert, Ed.). Edward Elgar.

Brock, W., & De Lima, P. (1996). Nonlinear Time Series, Complexity Theory, and Finance. In G. Maddala & C. Rao (Eds.), *Handbook of Statistics Volume 14: Statistical Methods in Finance*. Elsevier Science.

Brown, H. (2005). *Physical Relativity: Space-Time Structure from a Dynamical Perspective*. Oxford University Press.

Brown, H. (2020). Do Symmetries "Explain" Conservation Laws? The Modern Converse Noether Theorem vs. Pragmatism. In J. Read, B. Roberts, & N. Teh (Eds.), *The Physics and Philosophy of Noether's Theorems*. Cambridge University Press. Retrieved December 28, 2020, from https://arxiv.org/pdf/2010.10909.pdf

Brown, H., & Brading, K. (2002). General Covariance from the Perspective of Noether's Theorems. *Diálogos, 79*, 59–86.

Brown, H., & Myrvold, W. (2008). Boltzmann's H-theorem, Its Limitations, and the Birth of (Fully) Statistical Mechanics. Retrieved May 23, 2020, from https://arxiv.org/abs/0809.1304

Brown, H., & Read, J. (2019). The Dynamical Approach to Spacetime Theories. In E. Knox & A. Wilson (Eds.), *The Routledge Companion to Philosophy of Physics*. Routledge.

Bub, J. (2000). Indeterminacy and Entanglement: The Challenge of Quantum Mechanics. *The British Journal for the Philosophy of Science, 51*, 597.

Callender, C. (2016a). Thermodynamic Asymmetry in Time. *The Stanford Encyclopedia of Philosophy*, Winter 2016 ed. https://plato.stanford.edu/archives/win2016/entries/time-thermo/

Callender, C. (2016b). Thermodynamic Asymmetry in Time. In E. Zalta (Ed.), *The Stanford Encyclopedia of Philosophy.* Retrieved April 12, 2020, from https://plato.stanford.edu/archives/win2016/entries/time-thermo/

Calvin, J. (1813). *Institutes of the Christian Religion, Vol. I* (J. Allen, Trans.). Presbyterian Board of Publication.

Campbell, J., Lo, A., & MacKinlay, C. (1997). *The Econometrics of Financial Markets*. Princeton University Press.

Carroll, J. (2016). Laws of Nature. *The Stanford Encyclopedia of Philosophy.* Retrieved May 3, 2020, from https://plato.stanford.edu/archives/fall2016/entries/laws-of-nature/

Cartwright, N. (1983). Causal Laws and Effective Strategies. In *How The Laws of Physics Lie* (pp. 21–43). Oxford University Press.

Chatfield, C., & Xing, H. (2019). *The Analysis of Time Series*. Taylor & Francis.

Chiba, T., et al. (2007). Time-Variation of the Proton-Electron Mass Ratio and the Fine Structure Constant with a Runaway Dilation. *Physical Review D, 75*, 10.

Cicero, M. (1853). On Fate. In C. Yonge (Trans.), *The Treatises of M. T. Cicero* (pp. 264–282). Henry G. Bohn.

Cicero, M. (1989). On Fate. In H. Rackham (Trans.), *Rhetorical Treatises: Book III, De Fato v. 4: De Oratore*. Loeb.

Coinmarketcap.com data. (n.d.). Retrieved August 21, 2020, from https://coinmarketcap.com/charts/

Cramer, J. (1987). The Transactional Interpretation of Quantum Mechanics. *Reviews of Modern Physics, 58*, 647–688.

Cramer, J. (2015). The Transactional Interpretation of Quantum Mechanics and Quantum Nonlocality. Retrieved December 8, 2020, from https://arxiv.org/pdf/1503.00039.pdf

Dahlstrom, D. (2010). The *Critique of Pure Reason* and Continental Philosophy: Heidegger's Interpretation of Transcendental Imagination. In P. Guyer (Ed.), *The Cambridge Companion to Kant's Critique of Pure Reason*. Cambridge University Press.

Darby, G. (2010). Quantum Mechanics and Metaphysical Indeterminacy. *Australasian Journal of Philosophy, 88*(2), 227–245.

Davies, P. (2008). *The Goldilocks Enigma: Why Is the Universe Just Right for Life?* Houghton Mifflin.

Davies, P. (2010). Universe from Bit. In P. Davies & N. Gregersen (Eds.), *Information and the Nature of Reality* (pp. 65–91). Cambridge University Press.

Davies, P., Davis, T., & Lineweaver, C. (2002). Cosmology—Black Holes Constrain Varying Constants. *Nature, 418*, 602–603.

De Pierris, G., & Friedman, M. (2018). Kant and Hume on Causality. In E. Zalta (Ed.), *The Stanford Encyclopedia of Philosophy*. Retrieved March 7, 2021, from https://plato.stanford.edu/archives/win2018/entries/kant-hume-causality

Degl'Innocenti, S., et al. (1996). Time-Variation of Newton's Constant and the Age of Globular Clusters. *Astronomy and Astrophysics, 312*, 345–352.

Deleuze, G. (2004). *Difference and Repetition* (P. Patton, Trans.). Continuum.

DeWitt, B. (1970). Quantum Mechanics and Reality. *Physics Today, 23*(9), 30–35.

Dirac, P. (1930). *Principles of Quantum Mechanics*. Oxford University Press.

Dirac, P. (1937). The Cosmological Constants. *Nature, 139*, 323.

Dirac, P. (1938). A New Basis for Cosmology. *Proceedings of the Royal Society, 165*(921), 199–208.

Dirac, P. (1939). The Relation Between Mathematics and Physics. *Proceedings of the Royal Society (Edinburgh), 59*, 122–129.

Dorrien, G. (2004). *Imperial Designs: Neoconservatism and the New Pax Americana*. Routledge.

Dvali, G., & Zaldarriaga, M. (2002). Changing α with Time: Implications for Fifth-Force Type Experiments and Quintessence. *Physical Review Letters, 88*, 091303.

Dzuba, V., Flambaum, V., & Webb, J. (1999a). Calculations of the Relativistic Effects in Many-Electron Atoms and Space-Time Variation of Fundamental Constants. *Physical Review A, 59*(1), 230–237.

Dzuba, V., Flambaum, V., & Webb, J. (1999b). Space-Time Variation of Physical Constants and Relativistic Corrections in Atoms. *Physical Review Letters, 82*(5), 884.

Earman, J. (1974). An Attempt to Add a Little Direction to "The Problem of the Direction of Time". *Philosophy of Science, 41*(1), 15–47.

Earman, J. (1986). *A Primer on Determinism*. D. Reidel.
Earman, J. (2004a). Determinism: What We Have Learned and What We Still Don't Know. Retrieved April 10, 2020, from https://www.ucl.ac.uk/~uctytho/Determinism.pdf
Earman, J. (2004b). Laws, Symmetry and Symmetry Breaking: Invariance, Conservation Principles, and Objectivity. *Philosophy of Science, 71*(5), 1227–1241.
Earman, J. (2006). The "Past Hypothesis": Not Even False. *Studies in History and Philosophy of Modern Physics., 27*(3), 399–430.
Einstein, A. (1970). Autobiographical Notes. In A. Schlipp (Ed.), *Einstein: Philosopher-Scientist*. Open Court.
Einstein, A., & Minkowski, H. (1920). *The Principle of Relativity* (M. Saha & S. Bose, Trans.). University of Calcutta.
Elitzur, A., & Cohen, E. (2011). The Retrocausal Nature of Quantum Measurement Revealed by Partial and Weak Measurements. *AIP Conference Proceedings, 120*, 1408.
Esposito, E. (2011). *The Future of Futures: The Time of Money in Financing and Society* (E. Esposito, Trans.). Edward Elgar.
Esposito, E. (2018a). Predicted Uncertainty: Volatility Calculus and the Indeterminacy of the Future. In J. Beckert & R. Bronk (Eds.), *Uncertain Futures: Imaginaries, Narratives, and Calculation in the Economy* (pp. 219–235). Oxford University Press.
Esposito, E. (2018b). The Time of Money in Finance and U.S. Society. *Finance and Society, 4*(1), 15–25.
Eyal, I., & Sirer, E. (2013). *Majority Is Not Enough: Bitcoin Mining Is Vulnerable*. https://arxiv.org/abs/1311.0243
Fama, E. (1970). Efficient Capital Markets: A Review of Theory and Empirical Work. *The Journal of Finance, 25*(2), 383–417.
Faye, J. (2018). Backward Causation. In E. Zalta (Ed.), *The Stanford Encyclopedia of Philosophy*. Retrieved October 4, 2020, from https://plato.stanford.edu/archives/sum2018/entries/causation-backwards/
Feigl, H., & Brodbeck, M. (Eds.). (1953). *Readings in the Philosophy of Science*. Appleton-Century-Crofts.
Feyerabend, P. (1989). Realism and the Historicity of Knowledge. *Journal of Philosophy, 86*, 393–406.
Flambaum, V. (2001). Does the Fine Structure Constant Vary with Time and Distance? In *Atomic Physics 17: International Conference on Atomic Physics* (Vol. 551, pp. 86–99). American Institute of Physics.

Forester, T. (1853). *The Chronicle of Henry of Huntingdon: Comprising the History of England, from the Invasion of Julius Caesar to the Accession of Henry II*. Henry G. Bohn.

Friederich, S., & Evans, P. (2019). Retrocausality in Quantum Mechanics. In E. Zalta (Ed.), *The Stanford Encyclopedia of Philosophy*. Retrieved December 6, 2020, from https://plato.stanford.edu/entries/qm-retrocausality

Friedman, M. (1983). *Foundations of Space-Time Theories: Relativistic Physics and Philosophy of Science*. Princeton University Press.

Friedman, M. (1992). Causal Laws and the Foundations of Natural Science. In P. Guyer (Ed.), *The Cambridge Companion to Kant* (pp. 161–199). Cambridge University Press.

Frigg, R., Berkovitz, J., & Kronz, F. (2020). The Ergodic Hierarchy. In E. Zalta (Ed.), *The Stanford Encyclopedia of Philosophy*. Retrieved November 1, 2020, from https://plato.stanford.edu/archives/fall2020/entries/ergodic-hierarchy/

Frigg, R., & Werndl, C. (2010). Explaining Thermodynamic-Like Behaviour in Terms of Epsilon-Ergodicity. *Philosophy of Science, 78*, 628–652.

Fritzsch, H. (2008). The Fundamental Constants in Physics and Their Time Dependence. *Progress in Particle and Nuclear Physics., 621*(2), 329–342.

Galison, P. (2004). *Einstein's Clocks, Poincaré's Maps: Empires of Time*. Hodder and Stoughton.

Ghirardi, G., Rimini, A., & Weber, T. (1986). Unified Dynamics for Microscopic and Macroscopic Systems. *Physical Review D, 34*(2), 470–491.

Ginsborg, H. (1997). Lawfulness Without Laws: Kant on the Free Play of Imagination and Understanding. *Philosophical Topics, 25*(1), 37–81.

Godone, A., et al. (1993). New Experimental Limits to the Time Variations of $g_p(m_e/m_p)$ and α. *Physical Review Letters, 71*(15), 2364–2366.

Goranko, V., & Rumberg, A. (2020). Temporal Logic. In E. Zalta (Ed.), *The Stanford Encyclopedia of Philosophy*. Retrieved August 7, 2020, from https://plato.stanford.edu/archives/sum2020/entries/logic-temporal/

Grimmett, G., & Stirzaker, D. (2001). *Probability and Random Processes*. Oxford University Press.

Grøn, Ø., & Hervik, S. (2007). *Einstein's General Theory of Relativity: With Modern Applications in Cosmology*. Springer Science & Business Media.

Grünbaum, A. (1953). Some Remarks on Professor Ushenko's Interpretation of Causal Law. *The Journal of Philosophy, 50*(4), 115–120.

Guyer, P. (1980). Kant on Apperception and "A Priori" Synthesis. *American Philosophical Quarterly, 17*(3), 205–212.

Guyer, P. (1987). *Kant and the Claims of Knowledge*. Cambridge University Press.

Guyer, P. (2008). *Knowledge, Reason and Taste: Kant's Response to Hume*. Princeton University Press.

Guyer, P., & Walker, R. (1990). Kant's Conception of Empirical Law. *Proceedings of the Aristotelian Society, Supplementary Volumes, 64*, 221–258.

Hamilton, J. (1994). *Time Series Analysis*. Princeton University Press.

Harrigan, N., & Spekkens, R. (2010). Einstein, Incompleteness, and the Epistemic View of Quantum States. *Foundations of Physics, 40*, 125–157.

Hawley, K. (1998). Indeterminism and Indeterminacy. *Analysis, 58*(2), 101–106.

Hegel, G. (1958). *Philosophy of Right* (T. Knox, Trans.). Clarendon Press.

Hegel, G. (1995). *Lectures on the History of Philosophy, Vol. 3: Medieval and Modern Philoospphy* (E. Haldane & F. Simson, Trans.). University of Nebraska Press.

Heidegger, M. (1997). *Kant and the Problem of Metaphysics* (R. Taft, Trans.). Indiana University Press.

Hesse, M. (1962). *Forces and Fields: The Concept of Action at a Distance in the History of Physics*. Dover Publications.

Hoefer, C. (2016). Causal Determinism. In E. Zalta (Ed.), *The Stanford Encyclopedia of Philosophy*. Retrieved May 2, 2020, from https://plato.stanford.edu/archives/spr2016/entries/determinism-causal/

Hull, J. (2002). *Options, Futures and Other Derivatives* (5th ed.). Prentice Hall.

Hume, D. (1999). *An Enquiry Concerning Human Understanding* (T. Beauchamp, Ed.). Oxford University Press.

Hume, D. (2000). *A Treatise of Human Nature* (D. Norton & M. Norton, Eds.). Oxford University Press.

Husbands, C. (2020a). Esposito's Temporality of Finance: Endogeneity and Revisability in Derivative Transactions. *Finance and Society, 6*(2), 114–129.

Husbands, C. (2020b). *Klossowski's Semiotic of Intensity*. De Gruyter.

Jacobi, F. (2012). *Werke* (Vol. 3). Nabu Press.

Johnson, T. (2016). The Necessity of Multi-disciplinary Scholarship for Finance: On Ayache and Roffe. *Finance and Society, 2*(2), 189–204.

Kafatos, M., Roy, S., & Roy, M. (2005). Variation of Physical Constants, Redshift and the Arrow of Time. *Acta Physica Polonica, 36*(10), 3139–3161.

Kajita, M., & Moriwaki, Y. (2009). Proposed Detection of Variation in m_p/m_e Using a Vibrational Transitional Frequency of a CaH^+ Ion. *Journal of Physics B, 42*, 6.

Kanekar, N. (2008). Do Fundamental Constants Change With Time? *Modern Physics Letters A, 23*(32), 2711–2725.

Kangasharju, J. (2013). *Distributed Systems: Synchronisation.* University of Helsinki. Retrieved August 23, 2020, from https://www.cs.helsinki.fi/webfm_send/1232

Kannisto, T. (2017). Kant on the Necessity of Causal Relations. *Kant-Studien, 108(4)*, 495–516.

Kant, I. (1933). *Immanuel Kant's Critique of Pure Reason* (N. Smith, Trans.). Macmillan.

Kant, I. (1958). *Critique of Practical Reason* (L. Beck, Trans.). Bobbs-Merrill.

Kant, I. (1985). Metaphysical Foundations of Natural Science. In J. Ellington (Trans.), *Philosophy of Material Nature.* Hackett.

Kant, I. (1995). *Opus Postumum* (E. Förster and M. Rosen, Trans.). Cambridge University Press.

Kant, I. (2004). *Prolegomena to Any Future Metaphysics* (G. Hatfield, Ed.). Cambridge University Press

Kant, I. (2005). *Notes and Fragments* (P. Guyer, Ed. and C. Bowman, P. Guyer, & F. Rauscher, Trans.). Cambridge University Press.

Kant, I. (2007). *Critique of Judgement* (J. Meredith, Trans.). Oxford University Press

Karimi, K. (2010). *A Brief Introduction to Temporality and Causality.* Retrieved April 11, 2020, from https://arxiv.org/ftp/arxiv/papers/1007/1007.2449.pdf

Keynes, J. (1936). *The General Theory of Employment, Interest, and Money.* Macmillan. Retrieved June 8, 2020, from https://www.files.ethz.ch/isn/125515/1366_KeynesTheoryofEmployment.pdf

Kim, Y., et al. (1999). A Delayed Choice Quantum Eraser. *Physical Review Letters, 83*, 1–5.

Kitcher, P. (1982). Kant on Self-Identity. *The Philosophical Review, 91*(1), 41–72.

Kitcher, P. (1998). Projecting the Order of Nature. In P. Kitcher (Ed.), *Kant's Critique of Pure Reason: Critical Essays* (pp. 219–238). Rowman & Littlefield.

Kochen, S., & Specker, E. (1967). The Problem of Hidden Variables in Quantum Mechanics. *Journal of Mathematics and Mechanics, 17*, 59–87.

Koistinen, O. (2003). Spinoza's Proof of Necessitarianism. *Philosophy and Phenomenological Research, 67*(2), 283–310.

Kox, A., Klein, M., & Schulmann, R. (1997). *The Collected Papers of Albert Einstein, Volume 6, The Berlin Years: Writings, 1914–1917.* Princeton University Press.

Kozlov, M., et al. (2004). Space-Time Variation of the Fine-Structure Constant and Evolution of Isotope Abundances. *Physical Review, A, 70*, 6.

Kožnjak, B. (2015). Who Let the Demon Out? Laplace and Boscovich on Determinism. *Studies in the History and Philosophy of Science, 51*, 42–52.

Ladik, J. (2000). Did the Universe Evolve by Chance? Remarks to Lee Smolin's Book "The Life of the Cosmos". *Journal of Molecular Structure: Theochem, 500*(1–3), 1–3.

Lamport, L. (1978). Time, Clocks and the Ordering of Events in a Distributed System. *Communications of the ACM, 21*(7), 558–565.

Lamport, L. (1998). The Part-Time Parliament. *ACM Transactions on Computer Systems, 16*(2), 133–169.

Lamport, L. (2001). Paxos Made Simple. Retrieved August 24, 2020, from http://lamport.azurewebsites.net/pubs/paxos-simple.pdf

Lamport, L. (2005). Generalized Consensus and Paxos. Retrieved August 24, 2020, from https://www.microsoft.com/en-us/research/wp-content/uploads/2016/02/tr-2005-33.pdf

Land, N. (2018). *Crypto-Current, An Introduction to Bitcoin and Philosophy.* Retrieved August 22, 2020, from http://sumrevija.si/wp-content/uploads/2018/11/PDF.pdf

Landau, S., & Vucetich, H. (2002). Testing Theories That Predict Time Variation of Fundamental Constants. *Astrophysics Journal, 570*(2), 463–469.

Langacker, P. (2004). Time Variation of Fundamental Constants as a Probe of New Physics. *International Journal of Modern Physics, 19*, 157–165.

Lansky, J. (2018). Possible State Approaches to Cryptocurrencies. *Journal of Systems Integration, 9*(1), 19–31.

Laplace, P. (1902). *A Philosophical Essay on Probabilities* (F. Truscott & F. Emory, Trans.). John Wiley & Sons.

Leifer, M., & Pusey, M. (2017). Is a Time Symmetric Interpretation of Quantum Theory Possible Without Retrocausality? *Proceedings of the Royal Society A, 473*, 2202. Retrieved October 31, 2020, from https://arxiv.org/pdf/1607.07871.pdf

Lewis, P. (2013). Retrocausal Quantum Mechanics: Maudlin's Challenge Revisited. *Studies in History and Philosophy of Science Part B: Studies in History and Philosophy of Modern Physics, 44*(4), 442–449.

Lindsay, R., & Margenau, H. (1963). *Foundations of Physics*. Dover Publications.

Lipkind, D. (1979). Russell on the Notion of Cause. *The Canadian Journal of Philosophy, 9*(4), 701–720.

Livio, M., & Stiavelli, M. (1998). Does the Fine-Structure Constant Really Vary in Time? *The Astrophysics Journal, 507*, L13.

Lowe, E. (1994). Vague Identity and Quantum Indeterminacy. *Analysis, 54*(2), 110–114.

Luhmann, N. (1981). Temporalstrukturen des Handlungssystems. In *Soziologische Aufklärung; 3. Soziales System, Gesellschaft, Organisation*. Westdeutsche Verlag.

MacKenzie, D. (2006). *An Engine, Not a Camera: How Financial Models Shape Markets*. MIT Press.

Malik, S. (2014). The Ontology of Finance. In *Collapse VIII* (pp. 629–812). Urbanomic.

Margenau, H. (1977). *The Nature of Physical Reality: A Philosophy of Modern Physics*. Ox Bow Press.

Massa, C. (1993). A New Limit on Time Dependence of the Gravitation al Constant from Gravity Modified Quantum Electrodynamics. *Astrophysics and Space Science, 209*, 307–308.

Massumi, B. (2010). Perception Attack: Brief on War Time. *Theory and Event, 13*, 3.

Massumi, B. (2015). *Ontopower: War, Powers, and the State of Perception*. Duke University Press.

Matherne, S. (2014). Kant and the Art of Schematism. *Kantian Review, 19*(2), 181–205.

Maudlin, T. (2011). *Quantum Non-Locality and Relativity: Metaphysical Intimations of Modern Physics* (3rd ed.). John Wiley & Sons.

Mellor, D. (1995). *The Facts of Causation*. Routledge.

Melnick, A. (1973). *Kant's Analogies of Experience*. Chicago University Press.

Messina, J. (2017). Kant's Necessitation Account of Laws and the Nature of Natures. In *Kant and the Laws of Nature* (pp. 131–149). Cambridge University Press.

Mill, J. (1974). *The Collected Works of John Stuart Mill, Volume 7 & 8: A System of Logic, Ratiocinative & Inductive Part 1*. University of Toronto Press and Routledge and Kegan Paul.

Miller, J. (1971). The Principle of Causality: Tautology, Empirical Proposition or Imperative? *The Southwestern Journal of Philosophy., 2*(1/2), 73–82.

Mosquera, M., & Civitarese, O. (2011). Time Variation of the Fine Structure Constant and of the Higgs Vacuum Expectation Value on Cosmological Time Scales. *Astronomy and Astrophysics, 526*, 1–4.

Müller, T. (2015). Time and Determinism. *Journal of Philosophical Logic, 44*(6), 729–740.

Myrvold, W. (2018). Philosophical Issues in Quantum Theory. *The Stanford Encyclopedia of Philosophy*. Retrieved November 20, 2020, from https://plato.stanford.edu/archives/fall2018/entries/qt-issues/

Nakomoto, S. (2008a). *Bitcoin: A Peer-to-Peer Electronic Cash System.* https://bitcoin.org/bitcoin.pdf

Nakomoto, S. (2008b). *Re: Bitcoin P2P E-Cash Paper.* https://satoshi.nakamotoinstitute.org/emails/cryptography/11/

Narayanan, A., et al. (2015). *Bitcoin and Cryptocurrency Technologies (Draft).* Princeton University. Retrieved January 2018, from http://www.the-blockchain.com/docs/Princeton%20Bitcoin%20and%20Cryptocurrency%20Technologies%20Course.pdf

Newton, I. (1999). *The Principia: Mathematical Principles of Natural Philosophy* (I. Cohen & A. Whiteman, Trans.). University of California Press.

Nietzsche, F. (1967). *The Will to Power* (W. Kaufmann & R. Hollingdale, Trans.). Random House.

Norton, J. (1993). General Covariance and the Foundations of General Relativity: Eight Decades of Dispute. *Reports on Progress in Physics, 56*, 791–858.

Norton, J. (2003). Causation as Folk Science. *Philosopher's Imprint, 3*, 4. Retrieved May 10, 2020, from https://quod.lib.umich.edu/p/phimp/3521354.0003.004/1

Padmanabhan, T. (1993). *Structure Formation in the Universe.* Cambridge University Press.

Peirce, C. (1891). The Architecture of Theories. *The Monist, 1*(2), 161–176.

Peirce, C. (1982–1993). *Writings of Charles S. Peirce: A Chronological Edition, Volumes 1–5.* Indiana University.

Penrose, R. (1996). On Gravity's Role in Quantum State Reduction. *General Relativity and Gravitation, 28*, 581–600.

Pereboom, D. (2019). Kant's Transcendental Arguments. In E. Zalta (Ed.), *The Stanford Encyclopedia of Philosophy*. Retrieved March 21, 2021, from https://plato.stanford.edu/archives/spr2019/entries/kant-transcendental

Petersen, C. (2017). *Kant on Time: Self-affection and the Constitution of Objectivity in Transcendental Philosophy.* Doctoral Thesis. Retrieved February 13, 2021, from http://repository.essex.ac.uk/21752/

Pippin, R. (1987). Kant on the Spontaneity of Mind. *Canadian Journal of Philosophy, 17*(2), 449–476.

Plato. (2002). Meno. In G. Grube (Trans.), *Plato: Five Dialogues*. Hackett.

Plutarch. (1952). *De Stoicorum Repugnantiis* (M. Pohlenz & R. Westman, Eds.), VI, 2.

Poincaré, H. (1913). The Measure of Time. In *The Foundations of Science (The Value of Science)* (pp. 222–234). Science Press.

Pooley, O. (2017). Background Independence, Diffeomorphism Invariance, and the Meaning of Coordinates. In D. Lehmkuhl, G. Schiemann, & E. Scholz (Eds.), *Towards a Theory of Spacetime Theories* (pp. 105–144). Springer.

Popper, K. (1982). *The Open Universe: An Argument for Indeterminism*. Routledge.

Price, H. (1984). The Philosophy and Physics of Affecting the Past. *Synthese, 61*(3), 299–323.

Price, H. (1996). *Time's Arrow and Archimedes' Point: New Directions for the Physics of Time*. Oxford University Press.

Price, H. (2001). Backwards Causation, Hidden Variables and the Meaning of Completeness. *Pramana, 56*(2, 3), 199–209.

Price, H. (2008). Toy Models for Retrocausality. *Studies in History and Philosophy of Modern Physics, 39*(4), 752–761. Retrieved January 12, 2020, from https://arxiv.org/pdf/0802.3230.pdf

Price, H. (2012). Does Time-Symmetry Imply Retrocausality? How the Quantum World Says "Maybe". *Studies in the History and Philosophy of Modern Physics, 43*, 75–83.

Price, H., & Wharton, K. (2013). Dispelling the Quantum Spooks—a Clue that Einstein Missed? Retrieved November 11, 2020, from https://arxiv.org/pdf/1307.7744.pdf

Primas, H. (2002). Hidden Determinism, Probability and Time's Arrow. In H. Atmanspacher & R. Bishop (Eds.), *Between Chance and Choice. Interdisciplinary Perspectives on Determinism* (pp. 89–113). Imprint Academic.

Pusey, M., Barrett, J., & Rudolph, T. (2012). On the Reality of the Quantum State. *Nature Physics, 8*, 475–478. Retrieved November 11, 2020, from https://arxiv.org/pdf/1111.3328.pdf

Quine, W. (1961). On What There Is. In *From a Logical Point of View* (2nd ed., pp. 1–19). Harper and Row.

Richard, M. (1981). Temporalism and Eternalism. *Philosophical Studies: An International Journal for Philosophy in the Analytic Tradition, 39*(1), 1–13.

Robinson, J. (1981). *What Are the Questions? And Other Essays: Further Contributions to Modern Economics*. M. E. Sharpe.

Roffe, J. (2015). *Abstract Market Theory*. Palgrave Macmillan.

Russell, B. (1918). *Mysticism and Logic and Other Essays*. Longmans, Green and Co.

Russell, B. (1948). *Human Knowledge*. Simon and Schuster.

Ryckman, T. (2018). Early Philosophical Interpretations of General Relativity. In E. Zalta (Ed.), *The Stanford Encyclopedia of Philosophy*. Retrieved December 23, 2020, from https://plato.stanford.edu/entries/genrel-early/

Salmon, W. (1984). *Scientific Explanation and the Causal Structure of the World*. Princeton University Press.

Schaper, E. (1964). Kant's Schematism Reconsidered. *The Review of Metaphysics, 18*(2), 267–292.

Schmitt, E., & Stevenson, R. (2004). Admitting Intelligence Flaws, Bush Stands by Need for War. *New York Times*, July 10, A9.

Schopenhauer, A. (2018). *Sämtliche Werke*. Forgotten Books.

Sklar, L. (1993). *Physics and Chance*. Cambridge University Press.

Smolin, L. (1992). Did The Universe Evolve? *Classical and Quantum Gravity, 9*, 173–191.

Smolin, L. (1997). *The Life of the Cosmos*. Oxford University Press.

Smolin, L. (2013). *Time Reborn*. Penguin.

Soros, G. (1987). *The Alchemy of Finance: Reading the Mind of the Market*. Wiley.

Spekkens, R. (2005). Contextuality for Preparations, Transformations, and Unsharp Measurements. *Physical Review A, 71*, 052108.

Spinoza, B. (1988). Ethics. In E. Curley (Trans.), *The Collected Works of Spinoza: Volume I*. Princeton University Press.

Strawson, P. (1966). *The Bounds of Sense: An Essay on Kant's Critique of Pure Reason*. Methuen.

Suppes, P. (1970). *A Probabilistic Theory of Causality*. North Holland.

Teufel, T. (2017). Kant's Transcendental Principle of Purposiveness and the "Maxim of the Lawfulness of Empirical Laws". In M. Massimi & A. Breitenbach (Eds.), *Kant and the Laws of Nature*. Cambridge University Press.

Thomson, D. (1994). Jackknifing Multiple-WINDOW SPECTRA. *Proceedings of the IEEE International Conference on Acoustics, Speech and Signal Processing, VI*, 73–76.

Tolley, C. (2006). Kant and the Nature of Logical Laws. *Philosophical Topics, 34*, 371–407.

Trzęsicki, K. (2015). Indeterministic Temporal Logic. *Studies In Logic, Grammar and Rhetoric, 4*(1), 139–162.

Turner, B. (2007). The Paxos Family of Consensus Protocols. Retrieved August 24, 2020, from http://www.fractalscape.org/files/paxos-family.pdf

UBS Chief Investment Office. (2020). *UBS Global Real Estate Bubble Index 2020*. UBS Switzerland AG.

Ushenko, A. (1942). The Problem of Causal Inference. *Philosophy of Science, 9*(2), 132–138.

Ushenko, A. (1953). The Principles of Casuality. *The Journal of Philosophy, 50*(4), 85–101.

Uygar, A. (2020). *Kant on Laws*. Notre Dame Philosophical Reviews. Retrieved April 19, 2021, from https://ndpr.nd.edu/reviews/kant-on-laws/

Valaris, M. (2008). Inner Sense, Self-Affection and Temporal Consciousness. *Philosophers' Imprint, 8*(4), 1–18.

Van Fraassen, B. (1989). *Laws and Symmetry*. Clarendon Press.

Van Kampen, N. (1991). Determinism and Predictability. *Synthese, 89*(2), 273–281.

Varshalovich, D., & Potekhin, A. (1996). Have the Masses of Molecules Changed during the Lifetime of the Universe? *Astronomy Letters, 5*(1), 1–5.

Wallace, D. (2012). *The Emergent Multiverse: Quantum Theory According to the Everett Interpretation*. Oxford University Press.

Wallace, D. (2016). *What Is Orthodox Quantum Mechanics?* Retrieved October 9, 2020, from https://arxiv.org/pdf/1604.05973.pdf

Warren, D. (2017). Kant on Mathematical Force Laws. In *Kant and the Laws of Nature* (pp. 171–192). Cambridge University Press.

Watkins, E. (2005). *Kant and the Metaphysics of Causality*. Cambridge University Press.

Watkins, E. (2017). Kant's Necessitation Account of Laws and the Nature of Natures. In *Kant and the Laws of Nature* (pp. 11–29). Cambridge University Press.

Watkins, E. (2019). *Kant on Laws*. Cambridge University Press.

Werndl, C. (2009a). What Are the New Implications of Chaos for Unpredictability? *The British Journal for the Philosophy of Science, 60*(1), 195–220.

Werndl, C. (2009b). Deterministic Versus Indeterministic Descriptions: Not That Different After All? In A. Hieke & H. Leitgeb (Eds.), *Reduction, Abstraction, Analysis* (pp. 63–78). Ontos. Retrieved May 10, 2020, from http://charlottewerndl.net/DeterminismIndeterminism_WittgensteinSymposium08.pdf

Werndl, C. (2011). Observational Equivalence of Deterministic and Indeterministic Descriptions and the Role of Different Observations. In *EPSA Philosophy of Science: Amsterdam* (pp. 427–439). Springer.

Werndl, C. (2016). Determinism. In M. Griffith, T. Neil, & K. Timpe (Eds.), *Routledge Companion to Free Will.* Routledge. Retrieved May 10, 2020, from http://philsci-archive.pitt.edu/12166/1/DeterminismIndeterminismWordPittsburghArchiveWithF.pdf

Werndl, C., & Frigg, R. (2015). Reconceptualising Equilibrium in Boltzmannian Statistical Mechanics and Characterising its Existence. *Studies in History and Philosophy of Modern Physics, 49*, 19–31. Retrieved November 7, 2020, from https://arxiv.org/pdf/1510.02260.pdf

Wheeler, J. (1978). *Frontiers of Time*. University of Texas.

Wheeler, J., & Zurek, A. (1983). *Quantum Theory and Measurement.* Princeton University Press.

Whitehead, A. (1926). *Science and the Modern World: Lowell Lectures*. Cambridge University Press.

Whitehead, A. (1978). *Process and Reality: An Essay in Cosmology*. The Free Press.

Whitehead, A. (1985). *Process and Reality*. Free Press.

Wilson, J. (2013). A Determinable-Based Account of Metaphysical Indeterminacy. *Inquiry, 54*(4), 359.

Wittgenstein, L. (2012). *Tractatus Logico-Philosophicus* (C. Ogden, Trans.). Dover Publications.

Wolff, R. (1963). *Kant's Theory of Mental Activity: A Commentary on the Transcendental Analytic of the Critique of Pure Reason*. Harvard University Press.

Wood, G. (2018). *Ethereum: A Secure Decentralised Generalised Transaction Ledger.* https://ethereum.github.io/yellowpaper/paper.pdf

Woodward, J. (2019). Scientific Explanation. In E. Zalta (Ed.), *The Stanford Encyclopedia of Philosophy*. Retrieved May 2, 2020, from https://plato.stanford.edu/archives/win2019/entries/scientific-explanation/

Zilsel, E. (2003). The Genesis of the Concept of Physical Law. In *The Social Origins of Modern Science* (pp. 96–122). Kluwer Academic Publishers.

Index[1]

[1] Note: Page numbers followed by 'n' refer to notes.

C. Husbands, *The Temporality of Determinacy*,
https://doi.org/10.1007/978-3-030-86530-6

FSC
www.fsc.org
MIX
Papier | Fördert gute Waldnutzung
FSC® C083411

Kolibri 360 GmbH
Ferdinand-Jühlke-Straße 7
99095 Erfurt, Deutschland
produktsicherheit@kolibri360.de